The Temptations
of Evolutionary Ethics

The Temptations
of Evolutionary
Ethics

Paul Lawrence Farber

University of California Press

Berkeley / Los Angeles / London

University of California Press
Berkeley and Los Angeles, California

University of California Press
London, England

Library of Congress Cataloging-in-Publication Data

Farber, Paul Lawrence, 1944–
    The temptations of evolutionary ethics / Paul
Lawrence Farber.
        p.   cm.
    Includes bibliographical references and index.
    ISBN 0-520-08773-9 (hard)
    1. Ethics, Evolutionary—History.   I. Title.
BJ1298.F37   1994
171'.7—dc20
                                94-5507
                                    CIP

Printed in the United States of America

1 2 3 4 5 6 7 8 9

The paper used in this publication meets the minimum
requirements of American National Standard for Information Sciences—Permanence of Paper for Printed
Library Materials, ANSI Z39.48-1984 ⊗

In memory of
Charles Farber

One lesson at least has been taught us so forcibly by our historical and critical studies in the theory of Ethics that we ought never to forget it in future. This is the extreme complexity of the whole subject of human desire, emotion, and action; and the paradoxical position of man, half animal and half angel, completely at home in none of the mansions of his Father's house, too refined to be comfortable in the stables and too coarse to be at ease in the drawing room. So long as we bear this lesson in mind we can contemplate with a smile or a sigh the waxing and waning of each cheap and easy solution which is propounded for our admiration as the last word of "science." We know beforehand that it will be inadequate; and that it will try to disguise its inadequacy by ignoring some of the facts, by distorting others, and by that curious inability to distinguish between ingenious fancies and demonstrated truths which seems to be the besetting weakness of men of purely scientific training when he steps outside his laboratory. And we can amuse ourselves, if our tastes lie in that direction, by noticing which well-worn fallacy or old familiar inadequacy is characteristic of the latest gospel, and whether it is well or ill disguised in its new dress.

—C. D. Broad

Contents

Acknowledgments

My father, to whom this book is dedicated, used to make distinctions when I was young which puzzled me. He would say things like, "Not fair? Do you want to talk about the law or about equity?" Or, "Don't ask me why, that's the way it is." Like many of those who are discussed in this book, he knew right from wrong but was not sure why. His remarks set me thinking about ethics at an early age.

The research for this book could not have begun or been completed without the resources of major libraries. I was fortunate to have a few months to begin my research in the reading room of the British Library in Bloomsbury during the winter of 1983 while I was teaching in a Northwest Interinstitutional Council for Study Abroad program. This was followed by a sabbatical year in Cambridge in 1988–89 where I made use of the Cambridge University Library and the libraries of Trinity College, Robinson College, and the Department of the History and Philosophy of Science. In between, the Oregon State University library has supplied my needs and when not able to, has very efficiently obtained the items through interlibrary loans. The librarians in all of these institutions have been helpful, and I feel greatly indebted to them.

It would not have been possible to spend a year in Cambridge had it not been for a National Endowment for the Humanities (NEH) grant (RH-20853-88). Dan Jones of the NEH was helpful in guiding me through the process of applying.

I have been encouraged and helped by many individuals. Maureen Chandler, Marsha Richmond, Jim Secord, Keith Benson, Tom Franzel, Maggie Osler, Michael Ruse, Ernan McMullin, Julie Curtis, Mikulàš Teich, and Jon Franklin aided me in different ways. The Warden and Fellows of Robinson College, by extending to me a bye-fellowship, made my Cambridge stay very pleasant and profitable. Joseph Needham spent time answering questions, and Martha Morris spent time telling me things about which I had not asked. Karla Russell guided me through computer manuals and provided secretarial backup that can be described without exaggeration as optimum. Mike Mix and Keith King, with whom I was involved in another project, were patient and understanding with my schedule, and Fred Horne encouraged my work all along. As has been true for more than twenty-five years, Vreneli Farber was behind the scenes, this time single parenting for three months with twin kindergarteners and making it possible in many other ways for the work to get done.

Introduction

Roughly fifteen years ago an undergraduate came to see me for advice about graduate school. He was a bright, sincere student and elicited a sympathetic reaction from me. Like many students then, especially those who were reacting to the recent Vietnam War as well as to what appeared to be an imminent energy crisis, he was profoundly concerned about the state of the world. During our conversation he said that he thought we might be able to deal better with current problems if we were to have an understanding of the nature of man.[1] He went on to explain that his goal was to do a master's thesis on the nature of man because "nothing had been done on the subject."

I was, understandably, amused by his naiveté and appalled by his ignorance. How a student could reach the last year of college without ever having had a glimpse of the enormous body of literature on human nature said a great deal about the failure of our educational system. I gave the student a short list of major philosophers. When I finally saw him again more than a year later, he did not refer to the subject. (He is currently an accomplished historian.)

But I thought quite a few times of that conversation, for that student was on to something. Our picture of man is out of focus, not for lack of writing on the subject but rather for lack of a satisfying perspective. Our conversation reminded me not only that there was something wrong with our highly specialized and fragmented educational system, which had fostered his opinion concerning the deficient state of literature on human nature, but also that there was something problematic about the literature he had been reading. He was, I came to realize, reacting against popularizations of "sociobiology." In his view, those attempts to explain

---

1. He was using the term in a gender-free sense, as I shall be doing also.

human nature from an evolutionary perspective could not provide an adequate moral or political vision for the twentieth century. The issue is hardly a new one, for the attempt to understand the nature of man from an evolutionary point of view is as old as the theory itself, and scientists, philosophers, social scientists, and other writers have debated the problem for well over a century. During the last twenty years considerable attention has been paid to the issue. Many of the topics currently being discussed on the implications of evolution for humans, however, have been treated at great length in the past, and quite a few of the contemporary arguments are simply recycled versions of earlier ones.

For the historian, therefore, many of today's debates are extremely frustrating. Do we have to refute arguments that were decisively rejected fifty years ago? Is our ignorance of history really condemning us to repeat the mistakes of the past? It would seem so. We read daily in journals and magazines that biology holds the key to human nature. These claims go far beyond the accepted commonsense realization that since humans evolved like other animals, the sciences that investigate animal physiology, behavior, and psychology are valuable fields for providing basic information about ourselves that has to be considered in any overview of human nature. Rather, what is being asserted by numerous writers is that the theory of evolution provides the foundation for a deeper analysis of human action than anything available in the humanities or social sciences. Their position starts from the premise that because humans, like all other animals, have evolved, their physical and behavioral traits are understandable in evolutionary terms. Biologists who study the evolutionary significance of animal behavior have produced a large and impressive body of literature. Can those insights be of value in understanding human social interaction? Can biology help us to comprehend better individual human motivation? Human responsibilities?

At issue in applying scientific knowledge to aid us in constructing a political or moral vision is the question, to what extent can biology explain human culture? Our culture not only is the locus of our political and moral values; for many, it also constitutes what most dramatically distinguishes us from other animals. Those who want to utilize biological knowledge to investigate human nature point out that humans are not unique in possessing culture. Animal behaviorists have carefully documented the existence of animal culture. For example, in a celebrated case, scientists observed a monkey that had accidentally "discovered" that washing sweet potatoes improved their flavor; in a few years, younger members of her band were imitating the process. Other similar cases of "social learning" involve birds, such as the blue tit, and are well known and often cited.

Can studies of animal social interaction provide insights into the wellsprings of human social interaction? It is, after all, a very far road from washing sweet potatoes to writing Talmudic commentaries. And while it may make a lot of sense to attempt to understand the social learning of monkeys or blue tits in terms of the potential adaptive advantages that such actions may confer, or in terms of the evolutionary background of such actions, the extension of that reasoning to human culture is highly controversial.

For one hundred years, scholars, commentators, and professionals in the humanities and social sciences have pointed out that human culture is complex, varied, and significantly different from the "culture" of any other animal. But that does not deter many contemporary writers. Timothy Goldsmith, the Andrew W. Mellon Professor of Biology at Yale University, recently wrote that "the science of biology has reached a point in its own development where it can begin to address some of the perennial questions about what it means to be human, themes that have been the province of literature and philosophy for centuries and supplemented by the social sciences in more recent times."[2] Like others who expound this same view, he means that an evolutionary reading of human behavior will produce valuable insights that have eluded those in the humanities and social sciences. Is such optimism warranted?

Although biology has contributed to our understanding of the functioning of the human body, to recognizing more fully our ecological place in the biosphere, and, more recently, to preventing or curing many diseases, it has been singularly unsuccessful in solving social problems or providing moral guidance. Few people today find the extrapolations from studies on the Belding ground squirrel more suggestive about human nature and human society than the novels of Balzac. Moreover, when we examine past attempts to

2. Timothy H. Goldsmith, The Biological Roots of Human Nature: Forging Links between Evolution and Behavior (Oxford: Oxford University Press, 1991): x.

take lessons on how to live from nature, we see that they were often contrived arguments that merely read social values into nature. Even more disturbing, as historians have shown, are the disastrous consequences of such beliefs. The story of British, American, and German eugenics in the early part of this century shows the extent to which science has been used to codify social biases and preconceptions into policy. The horrors of the Nazi regime's racial hygiene policies stand as grim reminders of the extreme to which such a line of reasoning can go.

Modern biological scientists have distanced themselves from the early eugenics movement and have explicitly criticized its confusion of empirical science with social opinion. Given the hindsight of fifty years, it is relatively easy to see mistakes in the early eugenicists' program. What is more difficult is to delve into contemporary writings and uncover the hidden social assumptions that motivate or direct current opinion. Consider the Human Genome Project. Its supporters largely justify the enormous projected expense by its potential medical value. Critics, however, are skeptical for a couple reasons. They question the claim that increased knowledge of the genetic basis of various disorders will result in improved treatment. To date, very few therapies are available for known genetic diseases. The more significant issue, however, is not whether therapeutics will keep pace with diagnostics but what we as a society define as a medical problem. The concept of "genetic disease" is relatively new and has steadily been extended to encompass a wide array of conditions. Enthusiasts are convinced that we are on the brink of a revolution even more profound than the discovery of antibiotics in the first part of this century, for they believe that knowledge of the human genome will unlock the secrets of abnormal genes or abnormal genetic conditions. This information allegedly will provide insight into which genes control specific physical and behavioral traits as well as which genes control abnormal "predispositions"—everything from tendencies toward violence to susceptibility to viral infections. Skeptics worry not only about what appear to be vastly overblown hopes but also about how we define "abnormal." What exactly constitutes an abnormal condition, that is, a "genetic disease"? Are we unconsciously constructing a bogus vision that recalls the early eugenicists' blatant projections onto nature of social attitudes? In the first decades of the twentieth century, "feeblemindedness" was a trait that was thought to be inherited in a Mendelian fashion and to contribute to a significant number of social problems, from prostitution to criminality. With only the slimmest evidence, researchers today—followed by the popular press—have suggested that various behavioral traits, or tendencies to specific behaviors, have a genetic basis; antisocial behavior, depression, homosexuality, and schizophrenia have been argued to be coded in our genes. With a growing emphasis on the genetic basis of behavior and on defining "normal" and "abnormal" increasingly in medical terms, the potential for a replay of the mistakes of the early eugenics movement is not to be lightly dismissed.[3]

More and more one reads that the future evolution of humankind is in our hands, or, with future research, will be within our power to direct. What directions should be pursued? An important segment of the scientific community argues that it is better suited to define these directions than the self-appointed guardians of culture, the humanists and social scientists. E. O. Wilson, the author of Sociobiology, has been very explicit in "biologicizing" the problem of human destiny. It is from an evolutionary perspective, he and others argue, that we can most reliably uncover a deeper humanism than that which the descendants of the scholars in classical curricula have to offer.

It is not just research biologists and their popularizers who are looking to evolutionary biology for new perspectives on human nature. Carl Degler, one of America's foremost historians and one who has convincingly shown how philosophical and ideological beliefs have informed our notions of human nature, has recently written sympathetically of "the return of biology" and the possibility of a real "science of man" based on evolutionary insights.[4]

But does biology hold the key, the code, to human nature? The study of nature may provide insight into psychological constraints, the sources of emotional energy, and the dynamics of interpersonal interaction, but can it provide a guide for human action and human values? The historical record is not very encouraging.

3. For a well-stated discussion, see Evelyn Fox Keller, "Nature, Nurture, and the Human Genome Project," in Daniel J. Kevles and Leroy Hood, eds., The Code of Codes: Scientific and Social Issues in the Human Genome Project (Cambridge: Harvard University Press, 1992): 281–299.

4. Carl N. Degler, In Search of Human Nature: The Decline and Revival of Darwinism in American Social Thought (Oxford: Oxford University Press, 1991).

This book looks at a central chapter in the history of attempts to understand human nature from an evolutionary perspective: the use of evolution in the Anglo-American world as a foundation for ethics. As we shall see, that history comprised three episodes, the first commencing after the formulation of the theory of evolution and lasting until after the First World War. The theory of evolution developed at a critical time in the history of Western ethics. In the decade following Charles Darwin's publication of the Origin of Species (1859), philosophers, clergymen, scientists, literary figures, and other intellectuals faced the problem of trying to establish a new foundation for ethics. The age of the Industrial Revolution had totally altered the fabric of society, and the intellectual and moral responses to those multidimensional changes were profound. In the Anglo-American world many judged as inadequate the former religious foundation for morality, and an intense reexamination of the basis for human morality took place. Some argued for a continuation of the Christian worldview. Others argued for a reformed Christian perspective. Still others looked to science for a new foundation. The latter, in particular, hoped evolution theory might serve as a guide, for it provided a scientific account of the origin of current living forms. Might it not also tell us something about life's destiny? Might it shed light on the origin of the moral sentiment? Might it provide some guidance for the future?

Darwin himself addressed the issue of man's "moral sentiment" in his Descent of Man (1871). It was a topic that he had grappled with for several decades and that he believed was vital to a comprehensive theory of evolution. His main concern was to explain the origins of morality, and in his published writings he avoided using science to justify or to challenge accepted beliefs. Darwin's evolutionary approach attracted several writers, such as William Kingdon Clifford and Leslie Stephen, who attempted to extend his ideas. Their efforts consisted of fashioning evolutionary justifications for accepted moral principles, but they did so in such a transparently contrived fashion that their failure to gain acceptance is understandable.

There were others, however, who sought a more robust treatment of evolutionary moral philosophy and who contended that it could be a guide for proper living rather than just an abstract justification of accepted norms. Darwin's contemporary, Herbert Spencer, was the most important of these, and his views were of special relevance for the history of evolutionary ethics. Evolution was central to Spencer's entire philosophy, and in a number of works he elaborated on how his evolutionary philosophy provided a base for ethics and politics.

The Darwinian and Spencerian beginnings of evolutionary ethics spawned a great outpouring of writings. They were eagerly read by those who were searching for a new foundation for morality. Spencer, in particular, caught the attention of both secular and religious reformers. Neither he nor his followers nor Darwin's followers, however, were successful in convincing professional philosophers or in establishing a lasting base of support among the more serious reading public. Philosophers like Henry Sidgwick at Cambridge picked apart the arguments and revealed serious logical flaws and inconsistencies in evolutionary ethics. Professional biologists with impeccable evolutionary credentials such as Thomas Henry Huxley and Alfred Russel Wallace rejected the position. The popular audience, however, was more receptive. In spite of its chilly reception in the academy, parsonages, and private clubs, proponents of evolutionary ethics received a limited hearing among the general public. A number of writers who made use of an evolutionary perspective on social and political issues extended the life of evolutionary ethics by tying it to ambitious new social programs. These broad overviews sidestepped technical philosophical issues and concentrated instead on the alleged implications of evolution for social policy. Benjamin Kidd and Woods Hutchinson, for example, each had, for a time, a substantial audience. They were individuals whose overriding concern went far beyond evolutionary ethics. Each realized that he could construct an evolutionary argument to promote his social agenda. Their acceptance of evolutionary ethics was closely tied, therefore, to a more general set of global concerns and, to an extent, shielded it from direct attack.

But to little avail. By the end of the First World War, evolutionary ethics, along with the sweeping philosophical systems that proliferated at the turn of the century, were in serious decline and might have sunk into oblivion had not a few prominent biologists renewed the campaign. In so doing, they initiated a second episode in the history of evolutionary ethics. In part, the newly synthesized modern theory of evolution inspired them with the hope that questions formerly unanswered by Darwinian evolution could now be profitably broached. Equally important, excitement generated by the "new psychology" led them and others to believe that Freudian insights had opened up an understanding of the human mind that was previously unattainable. These stunning successes in the biological and human sciences gave hope to a set of scientists that the time was ripe for the elaboration of a new "scientific humanism" that would serve as a modern worldview. Authors like Julian Huxley and C. H. Waddington made extensive efforts to renew interest in evolutionary ethics as part of this new worldview, but their versions were soon found to be wanting for the old reasons. That is, the new approaches were neither convincing nor able to resolve the basic philosophical issues raised earlier.

By the 1960s, evolutionary ethics was philosophically marginal. Instead of trailing off into the fringes of academic discourse, however, a third episode in its history began in the 1970s as a result of a revision inspired by sociobiology. This new evolutionary ethics differed from earlier ones in being better grounded in biology and being more moderate in its claims. Instead of attempting to directly derive moral lessons from nature, sociobiologists often spoke of inherited tendencies and of the interaction between cultural and genetic factors. Moreover, the attempt to ground an understanding of morality in nature fit very well into a general revival in favor of biological explanation. Contemporary evolutionary ethics has to be seen against a backdrop of other attempts to view human nature in a biological framework. Government agencies and university budgets have devoted impressive funding to understanding the mind through "cognitive science" and neurobiology. Similarly, statistical correlations between social behavior and physical makeup are published in major journals regularly. The Human Genome Project has given impetus to the biological approach to human nature by stressing possible genetic factors that may influence behavior. Even in anthropology, which for years has consciously contrasted "cultural" factors with "biological" ones, some researchers in recent years have been going back to evolution as a starting point for theories of human cultural development.

Has the new interest in the biological view of human nature altered the climate of acceptance for evolutionary ethics? A few philosophers enthusiastically champion it, and a number of otherwise hostile philosophers admit that any knowledge of basic human drives or tendencies is important in an overall picture of human nature. A survey of the literature, however, reveals that once again evolutionary ethics has met a lack of success. Philosophers, as well as much of the reading public, are skeptical that values can be derived legitimately from evolutionary biology. If anything, the tilt, particularly in the academic world, seems to be more in favor of a view that humans read values into nature, rather than discover them there. Deconstructionists in literary criticism, postmodernists in history, and social constructivists in philosophy and sociology all emphasize the subjective and/or social origins of values.

The repeated attempts to construct an evolutionary ethics, even if not successful, is nonetheless a topic that is historically interesting. Is there something inherently suggestive in the evolutionary perspective that accounts for its refusal to move offstage? Have philosophers formulated the issues in ethics in such a way that evolution is banned but not discouraged? In spite of repeated failures of evolutionary ethics to gain acceptance, do evolutionary insights have some value for moral philosophy?

To answer such questions, a careful study of the history of evolutionary ethics in the context of the philosophical attempts to establish a foundation for ethics is necessary. But such a study is devilishly tricky. For one thing, evolutionary ethics has varied considerably over the years. Moreover, so has the theory of evolution. Worse, so have the interpretations of evolutionary theory. In what follows I discuss the major versions of evolutionary ethics from Darwin to the present and attempt to explain the differences among them. I have not described every variant, nor have I looked at the internal development of each author. A wealth of literature exists on most of the individuals discussed. I have tried instead to tell what I believe is the story of a debate that has continued for more than a century. I have concentrated on the Anglo-American tradition because it is a discrete story. The debates over evolutionary ethics in countries like France or Germany had little impact on the Anglo-American discussions. A comparative study would be very enlightening, but that will have to wait until scholars examine other national traditions. At the end of this book I suggest some of the lessons that I believe can be learned from it. Given the diversity of previous opinion, it will not be surprising if others reach different conclusions. I hope, however, that this history will supply a useful framework for considering the topic.

1. Charles Darwin

For Charles Darwin, 1838 was an extraordinary year. Since his return in 1836 from a four-year expedition around the world, he had been actively engaged in research on geology and natural history, which had paid off handsomely in scientific results and professional recognition. Unlike his undergraduate days when he showed little promise in pursuing a profession, his work was now focused and productive. His future no longer loomed as a problem for the family. Moreover, he was ready for the serious responsibility of married life, and his proposal had been accepted by his childhood friend, Emma Wedgwood. The match was highly approved, and through parental allowance, it brought financial independence. At twenty-nine, then, in spite of an inauspicious beginning, Darwin had arrived at the threshold of adult, middle-class respectability.

Like several Victorian bachelors, however, it was his secret life that held the most interest. For it was in 1838, on September 28, that Darwin formulated the key to understanding "that mystery of mysteries—the first appearance of new beings on this earth." He had been convinced for many months that species changed and that these transformations shed light on many of the facts of natural history. But until September he had not satisfactorily conceived of a mechanism that could explain such change. And it was a mechanism that he sought, one that could be understood in simple physical terms.[1]

It would be twenty years, however, before Darwin made this

1. The literature on Darwin and his formulation of the theory of evolution is enormous and ever-growing. It is part of what is aptly referred to as the "Darwin Industry." John C. Greene provided a useful survey of the field in his article "Reflections on the Progress of Darwin Studies," Journal of the History of Biology 8, no. 2 (1975): 243–273. More recent surveys are David Oldroyd, "How Did Darwin Arrive at His Theory? The Secondary Literature to 1982," History of Science 22 (1984): 325–374; and Antonello La Vergata, "Images of Darwin: A Historiographic Overview," in David Kohn, ed., The Darwinian Heritage (Princeton:

work known, and even then, he did so reluctantly. Darwin had numerous reasons for delaying public discussion of his ideas. Evolutionary hypotheses were in bad odor in the 1830s and 1840s
among the intellectual elite, for they were associated with radical
politics, vulgarized science for lay readers, and speculative French
and German writings.[2] But it was more than just the potential
damage to his newly acquired scientific standing that kept Darwin
from publishing. He realized evolutionary explanations were sufficiently revolutionary that to be accepted, he would have to demonstrate how they were compatible with the known facts of nature. The pioneers of the mechanical philosophy in the seventeenth
century, such as Descartes, had faced a similar problem. To com-
bat scholastic philosophy and Renaissance naturalism, they produced vast surveys of nature showing how the new philosophy
"saved the phenomena." They thereby hoped, if not to demonstrate, at least to show the potential potency of their worldview.

Darwin's generation was raised with the idea that God created
the world with perfectly adapted products. The position was one
that had characterized English natural history since the seventeenth century. William Paley eloquently elaborated this view in
the early nineteenth century in his Natural Theology (1802), a book
of enormous popularity and influence. By the late 1830s, however,

Princeton University Press, 1985). The Kohn volume reflects much of the best work
in the Darwin trade.

For the general reader, the following three books have considerable value: Peter
Bowler, Evolution: The History of an Idea (Berkeley, Los Angeles, and London:
University of California Press, 1984); Michael Ruse, The Darwinian Revolution
(Chicago: University of Chicago Press, 1979); and Adrian Desmond and James
Moore, The Life of a Tormented Evolutionist: Darwin (New York: Warner Books,
1991).

On a more technical level, the publication of Darwin's manuscripts and correspondence makes it possible for scholars around the world to follow Darwin's intellectual development at a fine level of resolution. Especially useful are Paul Barrett et al., Charles Darwin's Notebooks, 1836–1844 (Ithaca: Cornell University
Press, 1987); and Frederick Burkhardt and Sydney Smith, eds., The Correspondence of Charles Darwin (Cambridge: Cambridge University Press, 1985–). The
Darwin correspondence project has the added value of containing letters to, as well
as from, Charles Darwin.

2. For a sense of the social ramifications and associations of evolutionary ideas
in the early part of the nineteenth century, see Adrian Desmond, The Politics of
Evolution: Morphology, Medicine, and Reform in Radical London (Chicago: University of Chicago Press, 1989). Desmond and Moore stress the social significance
of evolutionary ideas as a motif in their excellent biography of Darwin, The Life of
a Tormented Evolutionist.

much of Paley's work was under attack. Comparative anatomists argued that animal form could be understood in its own terms, not only in its relation to function. Naturalists and geologists visualized the surface of the earth as having changed and pointed to the fossil record as evidence that animal and plant life had altered dramatically through the millennia. Yet in spite of considerable criticism of his depiction of a static world of perfect adaptation, Paley's general perspective that science was uncovering God's plan of creation retained a firm grip on the official science of the time. Darwin was proposing a different vantage point from which to view the living world. Like the geologists, who were explaining the changes on the surface of the globe without reference to divine plans or divine intervention, he sought natural explanations for observed regularities in the living world. Although Darwin was willing to conceive of a power that created the laws of nature, he did not see the value of relying on supernatural forces to explain the specific facts of nature. For what did references to a divine plan tell us? Very little. As Darwin later wrote in the first edition of the Origin of Species, "It is so easy to hide our ignorance under such expressions as the 'plan of creation,' 'unity of design,' &c., and to think that we give an explanation when we only restate a fact."[3]

Darwin was looking for a scientific explanation of the facts in natural history, and for him, a scientific explanation meant a physical understanding.[4] His early notebooks expressed impatience with earlier writers who contemplated evolution with fuzzy concepts like animals "willing to change" (what he took to be Lamarck's opinion). Darwin faced a formidable problem: how to account in simple physical terms for the marvelous complexity of organic form; the exquisite adaptation of organisms to the environment; and the striking regularities of distribution patterns, classification systems, and the fossil record. Even more challenging, how was he to account for man and his seemingly unique attributes?

3. Charles Darwin, On the Origin of Species, Or the Preservation of Favoured Races in the Struggle for Life (reprint of 1st ed., Cambridge: Harvard University Press, 1966): 481–482.
    4. Darwin, like many of his contemporaries such as Huxley and John Tyndall, did not subscribe to a philosophical or a metaphysical materialism but rather a commonsense notion that the physical world we see around us is the subject of science. Hypothetical spirits, essences, and so on, were not accepted as valid scientific concepts. See Ruth Barton, "John Tyndall, Pantheist: A Rereading of the Belfast Address," Osiris, 2d ser., 3 (1987): 111–134.

Darwin, of course, was not the first to encompass humans in the sweep of natural history. Eighteenth-century naturalists, such as Linnaeus and Buffon, included man in their classifications of animals. But they were very definite that a significant gulf divided humans from the rest of living nature. Later writers who proposed a continuity between man and the "brutes," for example, Lamarck, were outside the pale of scientific respectability in Great Britain. Darwin knew this only too well. A few years earlier, when he was a medical student in Edinburgh, he had benefited from his friendship with the Scottish anatomist and supporter of Lamarck, Robert Grant. Grant went to London in 1827 but by 1838 was ostracized by the establishment, in part, because of his Lamarckism.[5]

Darwin recognized, however, that in spite of the potential unpopularity of the view, a secular evolutionary account of life had to account for man. Explaining the physical body of Homo sapiens was not difficult; it posed no greater problem than that of any other advanced species. Human behavior, however, stood as an obstacle. Darwin had given considerable thought to animal behavior, which he believed was a biological trait the evolution of which could be explained like the evolution of other traits. His approach to the evolution of animal behavior was similar to his general strategy of accounting for morphological traits, that is, to show a gradation in nature, from simple to complex, as illustrative of how a complex structure (or behavior) could have come into being through successive and gradual modifications.

But what of man's "higher" faculties? Could he provide a convincing scenario showing the emergence of what many held were sentiments that had no analogue in the rest of the animal world? Darwin was well prepared to consider the issue, for he had read extensively on human nature and on what was then called "political economy." The subjects were ones that he often discussed in his family and social circle, and his interest in them had already had significant consequences. On September 28, 1838, while reading a well-known essay by the Rev. Thomas Malthus, he made

5. Adrian Desmond discusses the complexity of the issue in his "Robert E. Grant: The Social Predicament of a Pre-Darwinian Evolutionist," Journal of the History of Biology 17 (1984): 189–223, and his more extensive study, The Politics of Evolution. Desmond shows how extensive transformationist thought was in Britain during the 1830s, but he also shows that it was primarily held by people on the margins of the scientific establishment. Desmond and Moore (1991) examine Darwin's relations with Grant.

an analogy between the author's account of mankind's population growth inevitably exceeding its food supply and what he knew of natural animal populations. Darwin quickly grasped that he could construct the solution to his general problem of formulating the mechanism of organic evolution.[6] It was this mechanism, natural selection, that he used in his Origin of Species to demonstrate how an evolutionary perspective could explain the facts of natural history. He constructed with it a new interpretation for the fields of classification, paleontology, biogeography, and comparative anatomy. What about human nature? Although Darwin believed human behavior could be explained by his new theory, he skirted the problem in the Origin of Species. He returned to it, several years later, in The Descent of Man, and Selection in Relation to Sex.

By 1871, however, the topic of human evolution was no longer novel. Thomas Henry Huxley, Alfred Russel Wallace, and Ernst Haeckel published works in the 1860s on human origins.[7] Others extended evolutionary ideas to social and moral questions. Francis Galton, Darwin's cousin, for example, published a widely read article in Macmillan's Magazine that discussed the heredity of mental and moral characteristics.[8] Galton's concern was in promoting eugenics, but in so doing, his article drew attention to the idea that natural selection could operate on psychological and moral traits. The essays in 1867 and 1868 by Walter Bagehot in the Fortnightly Review, which later were published as Physics and Politics, or Thoughts on the Application of the Principles of "Natural Selection" and "Inheritance" to Political Society (1872), interpreted

6. See Sandra Herbert, "The Place of Man in the Development of Darwin's Theory of Transmutation," Journal of the History of Biology, 2 pts., 7, no. 2 (1974): 217–258 and 10, no. 2 (1977): 155–227, for an interesting discussion of the wider issues of Darwin's thoughts on the evolution of humans. Scott Gordon questions the depth of Darwin's reading in political economy; see his "Darwin and Political Economy: The Connection Reconsidered," Journal of the History of Biology 22, no. 3 (1989): 437–459. However, more recently Desmond and Moore (1991) make a good case for his interest in and knowledge of writers such as Malthus.

7. Huxley published his Evidence as to Man's Place in Nature in 1863, Wallace's article "The Origin of Human Races and the Antiquity of Man Deduced from the Theory of 'Natural Selection'" appeared in 1864, and Haeckel's book, Natürliche Schöpfungsgeschichte, appeared in 1868. This latter work was translated and widely read in both the United States and Britain.

8. Francis Galton, "Hereditary Talent and Character," Macmillan's Magazine 12 (1865): 157–166 and 318–327.

human history in evolutionary terms and stressed the selective advantages of certain "virtues" like obedience, honesty, and valor.

In the Descent of Man, Darwin covered many of the standard topics concerning human evolution. Of particular interest to us, he addressed the issue of man's moral faculty, for it traditionally stood as a chasm between man and the rest of the organic world. Although Darwin wrote, "I fully subscribe to the judgment of those writers who maintain that of all the differences between man and the lower animals, the moral sense or conscience is by far the most important,"[9] he argued that the moral sense was amenable to investigation from the perspective of natural history. Where philosophers had merely speculated, he proposed a more profitable path of investigation. Man's moral sense was to be viewed as a problem of natural history, similar to the adaptation of the giant sloth to its environment or the origin of the structures of orchid blossoms. He explicitly stressed that the moral sense of man did not represent a unique characteristic; rather, it was a natural development for an intelligent social animal. Indeed, "any animal whatever, endowed with well-marked social instincts, would inevitably acquire a moral sense or conscience, as soon as its intellectual powers had become as well developed, or nearly as well developed, as in man."[10] It would probably be a different moral sense, but nonetheless a moral sense. Darwin did not mean to imply that any currently known animal other than man had a moral sense or that he questioned the statement that the moral sentiment was the best and highest distinction between man and the apes. Rather, what he wished to convey was that the difference was one of degree, not kind, and therefore did not serve to place man in a category forever removed from the lower animals.[11]

Human behavior, then, was a phenomenon to be explained scientifically. Although this view was in keeping with British empiricist philosophy, it broke with the optimistic tradition of natural theology that was firmly entrenched in official science. Instead, it advocated a less comfortable view of nature. In one of his early

---

9. Charles Darwin, The Descent of Man, and Selection in Relation to Sex (London: John Murray, 1871), 1: 70.

10. Ibid., 71–72.

11. For a detailed discussion of the development of Darwin's ideas on animal behavior, mind, and morals, see Robert Richards's Darwin and the Emergence of Evolutionary Theories of Mind and Behavior (Chicago: University of Chicago Press, 1987).

notebooks Darwin noted, "It is difficult to believe in the dreadful but quiet war of organic beings going on [in] the peaceful woods & smiling fields."[12] But he did not despair. In spite of the cycles of recession—indeed, the depression of 1837–1842 was the worst of the century—Englishmen of Darwin's social class in the 1830s were convinced that the great changes of the last half century (what we today call the Industrial Revolution) were bringing about an improvement in man's lot. They also stressed the ability of humans to improve. It was this belief in progress that took the sting out of evolution and for that matter, out of the enormous social dislocation that was taking place. Not even documents like Edwin Chadwick's Report of an Inquiry into the Sanitary Conditions of the Labouring Population of Great Britain (1842), which relentlessly recorded the filth, pollution, and disease among the working population, could dent the confidence of Darwin's social class. Herbert Spencer, whose ideas will be discussed in chapter 3, epitomized this optimistic perspective.

Darwin explained the progress of man in the same manner in which he explained the progress of life.[13] That is, the evolution of life and the evolution of culture were each the unanticipated result of a blind process of selection. Progress in the evolution of life was toward greater complexity; in human evolution, it was toward the emergence of higher cultures and civilizations. Progress was not the working out of a cosmic plan but rather the natural consequence of selection. This denial of a goal-directed view of evolution put Darwin in opposition to the majority of evolutionists of the nineteenth century, particularly those like Spencer who were primarily interested in human evolution.

It also led Darwin to give a different interpretation to traditional as well as contemporary ideas in moral philosophy and psychology. British empirical philosophers had long assumed that the workings of the mind were open to scientific investigation, and British moral philosophers, such as Adam Smith, developed ideas about a natural "sympathy" that was fundamental in understand-

---

12. Barrett et al., Charles Darwin's Notebooks, 429.
13. For a good discussion of Darwin's "Social Darwinism," see John C. Greene, "Darwin as a Social Evolutionist," Journal of the History of Biology 10 (1977): 1–27, also reprinted in his collection of essays, Science, Ideology, and World View: Essays in the History of Evolutionary Ideas (Berkeley, Los Angeles, and London: University of California Press, 1981).

ing the moral sentiment. Our natural sympathetic understanding of the reactions of others and the ability to put ourselves in another's place made it possible, according to this perspective, to derive happiness from and judge the actions of our fellowman. Although most British moral philosophers credited the Creator with implanting moral sentiments, the position was open to secular interpretations.

In Darwin's day the concept of moral sentiment had been reformulated by contemporary writers such as Alexander Bain, who drew on the utilitarian and associationist traditions in philosophy, as exemplified by John Stuart Mill, and supported it with the current physiological understanding of the brain and nervous system. Moral habits, according to this view, were accounted for by "sympathy" and the internalization of society's punishment of deviant behavior.[14] "Punishment, or the deliberate infliction of pain, in the name of the collective mass of beings making a society, is the foremost incentive to Duty, considered as abstinence from injuring others. . . . Hence duty is the line chalked out by public authority, or law, and indicated by penalty or punishment," Bain had written in a popular work of 1868.[15]

For Darwin the position was flawed. Bain, like Mill and similar writers, approached the origin of moral sentiments as a question of how individuals acquire moral opinions. From this perspective, individuals learned to be moral from the associations of ideas in their earliest personal experiences, molded by their family backgrounds and their social circumstances. Society judged the value of an act by reference to its results: that which produced the greatest happiness for the greatest number was good. Although Darwin found congenial much in the overall strategy of this utilitarian position, he agreed with its critics that it was not adequate. Why should happiness be equated with good? How can we account for

---

14. See Edward Manier, The Young Darwin and His Cultural Circle (Dordrecht: D. Reidel, 1978), and Richards, Darwin and the Emergence of Evolutionary Theories, for analyses of the development of Darwin's thought and his reading of the British moralists. J. B. Schneewind, Sidgwick's Ethics and Victorian Moral Philosophy (Oxford: Oxford University Press, 1977), has a useful and concise history of the British moral philosophy relevant to understanding Darwin.

15. Alexander Bain, Mental Science: A Compendium of Psychology, and the History of Philosophy (New York: D. Appleton, 1868): 393–394. The original English edition was published by Longmans, Green the same year and was entitled Mental and Moral Science.

such obvious discrepancies between theory and practice as the stronger feeling of obligation to friends and relatives than to strangers? Why this "natural" preference for those of close relation?

The flawed formulation was thrown into even greater relief by the ethnological literature of the time that dealt with morals.[16] The study of "primitive" peoples suggested two major conclusions. The first was the self-congratulatory one that European civilization was the most advanced moral state of man and that there existed a continuum from rude barbarism to high civilization to be found both in time and space; that is, contemporary societies exhibited a range of sophistication, and the historical record suggested a parallel. This was, of course, analogous to Darwin's view that the range of variation in organ systems showed a gradation from simple to complex which paralleled their historical record.

The second conclusion that "emerged" from the data was that although individual societies differed radically one from another and particular moral notions were relative to the society from which they arose (and could not be transposed validly), nonetheless, analysis revealed a set of universal ethical norms.

A natural ethics, therefore, was possible. But a naturalistic ethics, along the lines sketched by Bain or Mill, was not sufficient. Individual happiness, or even composite happiness, could not explain or justify moral sentiments. Darwin had read a number of moral writings in the late 1830s which described moral faculties as inborn, and these may have reinforced his dissatisfaction with utilitarian ethics. Sir James Mackintosh, for example, in his Dissertation on the Progress of Ethical Philosophy, Chiefly during the Seventeenth and Eighteenth Centuries (1836) described the moral sentiment as "educed by intercourse with the external world" but nonetheless "a law of our nature." And that great character of the London intellectual scene, Harriet Martineau, who popularized Malthus (and carried on an extended liaison with Darwin's brother, Erasmus), wrote in her relativistic How to Observe: Manners and Morals (1838) that although a comparison of different cultures

16. The Victorians were very interested in ethnology. A classic work on the topic is J. W. Burrow, Evolution and Society: A Study in Victorian Social Theory (Cambridge: Cambridge University Press, 1966). A review article by Gay Weber enlarges on the subject; see his "Science and Society in Nineteenth-Century Anthropology," History of Science 11 (1974): 260–283. The recent work of George Stocking, Jr., takes the story even further; see his Victorian Anthropology (New York: Free Press, 1987).

showed that what were regarded as vices or virtues were relative to their context and not absolute, nonetheless, all humans displayed some feelings of right and wrong. What were the origins of those intuitions, and how were they justified?

According to Darwin, an evolutionary perspective provided a solution to these dilemmas and potentially a more powerful account than any current ones. Philosophical, anthropological, and psychological viewpoints did not supply a sufficient foundation for ethics, but instead they constituted valuable raw material to be incorporated into a wider account. In the Descent of Man, Darwin sketched an evolutionary explanation that he believed more satisfactorily traced "the development of the intellectual and moral faculties."

Natural selection among individuals was thought to be a major force responsible for man's early intellectual progress. With greater intelligence came the possibility of the sentiments of sympathy, fidelity, and courage, which could give advantages to those groups that had individuals possessing them. Finally, there would result "a highly complex sentiment, having its first origin in the social instincts, largely guided by the approbation of our fellow-men, ruled by reason, self-interest, and in later times by deep religious feelings, confirmed by instruction and habit, [which,] all combined, constitute our moral sense and conscience."[17] A moral sense would likely be of little selective value to the individual; indeed, it might prove harmful. Darwin suggested, however, that a group possessing such individuals would have a significant adaptive advantage. Tribes having members who were willing to risk their own lives for the good of the community would supplant other tribes that lacked such altruistic men. This natural selection of groups, not individuals, served to extend man's moral development. The general advance of man was like the evolution of organic life, not predetermined and with retrogressive as well as progressive moves. But the historical record, like the paleontological record, revealed progress.

The picture, then, was one that was basically positive. Man had evolved like other animals. Mental evolution had given rise to higher faculties. Among these were moral sensibilities that although differing in particulars from culture to culture, nonetheless

17. Darwin, Descent of Man, 1: 165–166.

had a common foundation that could best be understood in evolutionary terms. What defined a moral action was its contribution to the "general good." The general good was not knowable through a God-given intuition, nor was it action that would promote the "greatest happiness." Instead, it was to be conceived in the light of evolution: what contributed to the general good of the community.

Darwin had qualms about reducing all of ethics to evolutionary terms. He recognized that higher civilizations gave rise to humanitarian impulses and concepts of universal brotherhood that were not adaptive. He also recognized that "the strangest customs and superstitions, in complete opposition to the true welfare and happiness of mankind, have become all-powerful throughout the world."[18] The overall emphasis in Darwin's writings on the moral sentiment, however, was on an evolutionary perspective.

Darwin's work on the origin of the moral sentiment was a major source for what later became known as the ethics of evolution, or evolutionary ethics. As will be seen, it was not the only source, and others (even if called "Darwinian") differed substantially from his. Darwinian ethics, as we might term Darwin's views, started from the basic assumption that the world could be understood in physical terms. It pictured an evolution of life from simple organisms through man as having come about by natural processes with natural selection as the primary force. It stressed continuity and gradual change over discontinuity. Man's behavior was part of natural history, as was the behavior of animals. Social organization was part of the evolutionary story, and man's particular mental development was a continuation of the social and mental organization that could be seen in the nonhuman realm. The moral sentiment developed from social instincts and gave certain groups a selective advantage. What was "good" or "bad" was relative to individual societies, as were notions of "right" or "wrong." But underneath the particulars of specific moral codes, there existed a set of basic and universal norms that rested on our experience as a social species and that could be validated by reference to their adaptive value. Although the most refined moral sensibilities that were a product of high civilizations went beyond simple adaptive considerations, Darwin held that the history of ethics was in the main an evolutionary one. It was a subject that interested him,

18. Ibid., 99.

and his reading and private notes suggest that he devoted a significant amount of time to it. It was not a topic, however, that he pursued in print after the Descent of Man. Having shown a plausible origin for the moral sentiment and having established its adaptive value, he believed that he had accomplished his aim of encompassing man within his theory of evolution.

Darwin may have been satisfied with his contribution to the subject, but there were problems with his view. If one accepted the basic evolutionary thrust of his perspective, how was it possible to justify those noble sentiments of civilized (i.e., British) man that were not adaptive? How were they, in fact, to be distinguished from the "strange customs and superstitions" of the Hindu? Darwin may have proposed a reasonable account of the origin of the moral sentiment, but by admitting that some of the highest moral sentiments were nonadaptive, he showed that evolution was not sufficient as a justification for the existence of these sentiments. On what grounds did Darwin base his opinion that those noble sentiments of civilized man were superior to the "absurd religious beliefs" of savages? An ethics that could not account for accepted moral values or serve as a moral guide hardly qualified as an adequate new position. On a more fundamental level, did survival of the group justify morality at all? Evolution might provide an explanation of the origin of the moral sentiment and might even be able to trace its development in broad outline from savage times to the present. But was such an account anything more than a record of useful opinion? Was survival, or "general good" to the community, a sufficient criterion to define moral value? Why was it more adequate than "happiness" or God's will?

Darwin believed that by approaching ethics from the perspective of natural history he had moved the discussion to a more satisfactory forum. Although he did not attempt to resolve difficulties in his view, or to work out a fully articulated evolutionary ethics, there were several people, taking their inspiration from him, who took up the challenge. Among the most well known were William Kingdon Clifford, who stressed adaptation as a criterion for moral action, and Leslie Stephen, who used evolution as a base on which to ground established moral conviction. Chapter 2 examines their views.

2. Darwinian Ethics

The controversy over Darwin's ideas was among the major scientific issues of the nineteenth century. Numerous historians have noted, however, that concern with the implications of the theory was what chiefly led to the most heated interchanges and was what attracted the most attention. Whether or not ancient populations of armadillos were transformed into the species that currently inhabit the new world was certainly a topic about which zoologists could strongly disagree. But it was in discussing the broader implications of the theory's interpretation—such as the rejection of a teleological view of nature, the attack on natural theology, or the depiction of man as merely an advanced ape—that tempers flared and statements were made which could transform what otherwise would have been a quiet scholarly meeting into a social scandal. According to one account, a woman fainted and had to be carried out after the famous interchange between Samuel Wilberforce and Thomas Henry Huxley at the 1860 Oxford meeting of the British Association for the Advancement of Science. It was not the debate over the biological adequacy of Darwin's theory that the lady found so shocking but Huxley's insult to the bishop occasioned by the latter's reference to the former's alleged simian ancestry.

With such explosive tinder, it is little wonder that Darwin's treatment of ethics, like his discussion of evolution, moved off in several directions. As will be seen in this and subsequent chapters, it was extended, incorporated, co-opted, opposed, and altered, but rarely ignored, by those who were interested in the nature of man and the foundations of the moral sentiment.

William Kingdon Clifford

One of the first to discuss the ethical implications of Darwin's work was William Kingdon Clifford. He had gone to Cambridge as an undergraduate in 1863 to study mathematics, but far from confining himself to a narrow preparation for the Tripos, he indulged "his native bent for independent reading and research going far beyond the subjects of the examination."[1] His rooms were the meetingplace for a wide circle of friends, and he actively participated in numerous intellectual groups at Cambridge. He was, for example, a member of the Grote Club, a small discussion group that grew out of after-dinner talks at the vicarage of the Rev. John Grote in Trumpington near Cambridge. Grote was the Knightbridge Professor of Moral Philosophy, and the original group attracted some of the brightest young minds in the university, including Henry Sidgwick, a future holder of the Knightbridge professorship. Clifford dazzled the group, even if for some "he was too fond of astonishing people."[2]

Clifford's brilliance did not go unnoticed, and in 1868 he was elected a fellow of Trinity. His appointment to a professorship at University College took him to London in 1871, and he was elected a fellow of the Royal Society in 1874. Clifford was an able speaker and continued to be a daring thinker. A mathematician, he was among the first to appreciate the philosophical importance of the revolutionary work done by Riemann and Lobachevski in non-Euclidian geometry.

It was at Cambridge that Clifford developed an interest in Darwin, and he emerged as the leader of a group that was "carried away by a wave of Darwinian enthusiasm."[3] For Clifford, Darwin provided a revolutionary perspective, which, like the non-Euclidian geometry that shattered the belief in an eternal and fixed geometry, undermined traditional beliefs. Frederick Pollock, the jurist, later recalled,

> We seemed to ride triumphant on an ocean of new life and boundless possibilities. Natural Selection was to be the master-key of the universe; we expected it to solve all riddles and reconcile all contradictions. Among other things it was to give us a new system of ethics, combining the exactness of the utilitarian with the poetical ideals of the transcendentalist. We were not only to believe joyfully in the

1. William Kingdon Clifford, Lectures and Essays by the Late William Kingdon Clifford, ed. Leslie Stephen and Frederick Pollock (London: Macmillan, 1879), 1: 12.

2. John Maynard Keynes, Essays in Biography (London: Macmillan, 1933): 172.

3. Clifford, Lectures and Essays, 1: 33.

survival of the fittest, but to take an active and conscious part in making ourselves fitter. At one time Clifford held that it was worth our while to practise variation of set purpose; not only to avoid being the slaves of custom, but to eschew fixed habits of every kind, and to try the greatest possible number of experiments in living to increase the chances of a really valuable one occurring and being selected for preservation.[4]

Clifford unfortunately died in 1879 at thirty-five. His lectures and essays were collected and published the year of his death by two of his friends, Leslie Stephen and Frederick Pollock. Several of the essays, which continued to be quoted throughout the century, treated ethics and had been published earlier in widely read journals like the Fortnightly Review and the Contemporary Review. In them he elaborated on Darwin's ideas as published in the Descent of Man. The moral sense, according to Clifford, was the product of evolution and was an adaptive mechanism for group survival. Clifford extended this view to establish a foundation for a naturalistic ethics. He argued that right and wrong could be defined by social efficiency. "Right is an affair of the community, and must not be referred to anything else."[5] Individual happiness, or individual preservation, was not as important as community good in determining moral right. "The first principle of natural ethics, then, is the sole and supreme allegiance of conscience to the community," he wrote in 1875.[6]

Clifford aspired to develop a complete treatment of ethics. He believed that such a project was feasible, and he gave considerable thought to an overall synthetic philosophy. In this, he was impressed by the work of Herbert Spencer (chap. 3). He was, however, critical of many of the speculative and poorly reasoned philosophical systems of the day. Of one singularly uninspiring attempt he said, "He is writing a book on metaphysics, and is really cut out for it; the clearness with which he thinks he understands things and his total inability to express what little he knows will make

4. Ibid.
5. Ibid., 2: 171. The quotation is from his essay "Right and Wrong: The Scientific Ground of Their Distinction," which first appeared in the Fortnightly Review in 1875.
6. Ibid., 172.

his fortune as a philosopher."[7] In contrast, Clifford's statements were clear and strikingly phrased. Time, however, did not permit him a chance to present a systematic view of the subject or attempt a broad account of moral behavior from an evolutionary point of view. But there were others who did.

Clifford's interest in a new foundation for ethics and a new philosophical perspective from which to view society reflected more than his own personal tastes. The 1870s were a period during which intellectuals reexamined their cultural assumptions. This "crisis of conscience," as it has been called by historians, had been brought about by numerous factors, mostly associated with the social and intellectual effects of the Industrial Revolution. Two major conceptual issues, in particular, precipitated a widespread sense that the received worldview was incompatible with contemporary ideas. One was the fruit of several decades of brilliant work in science, which suggested to many a new model for intellectual understanding, and the second was an equally important set of studies called the "higher criticism," which treated Scripture as a literary text rather than as a divine document. The combination of natural science and Bible criticism undermined the entire framework of Christian thought and morality. Even for the supporters of such a momentous change like Clifford, there was considerable ambiguity. For many of those who were willing to eradicate Christianity as a force in the modern world,[8] the loss of Christian morality was perceived to be an intolerable impoverishment of humanity. Much of the creative thought of the late nineteenth century was directed in one form or another at attempting to resolve this dilemma.[9] For some it suggested a need to return to

7. Ibid., 1: 28.
8. Clifford, for example, wrote, "I can find no evidence that seriously militates against the rule that the priest is at all times and in all places the enemy of all men." Clifford, Lectures and Essays, 2: 237.
9. Owen Chadwick has written perceptively on this issue. See especially his Secularization of the European Mind in the Nineteenth Century (Cambridge: Cambridge University Press, 1975), and The Victorian Church, 2 vols. (London: Adam and Charles Black, 1966–1970). Among the many other excellent studies, see M. A. Crowther, Church Embattled: Religious Controversy in Mid-Victorian England (Newton Abbot: David and Charles, 1970), and Howard Murphy, "The Ethical Revolt Against Christian Orthodoxy in Early Victorian England," American Historical Review 60, no. 4 (1955): 800–817, which provide some other dimensions of the issue.

orthodoxy, either Anglican or Roman. Others, following T. H. Green, sought a solution in modifying German idealism to modernize the Christian worldview. Still others, like Frederic Harrison, felt that Christianity should be abandoned and replaced with a "positive philosophy" as sketched by the French philosopher Auguste Comte.

Darwin's writings on the origin of the moral sentiment were suggestive to a wide set of intellectuals who looked to rationalism and science. What they found in their reading of Darwin was an important component of an alternative worldview, a naturalistic account of human origins and human faculties. Their quest was for a new foundation from which they could justify the values of contemporary society. The problem with their program, however, was their conflation of accounts of the origins of values with philosophical justifications of those values. Darwin succeeded in demonstrating to many the relevance of natural history for philosophy, but he had not provided a systematic account of ethics or a detailed defense for using evolutionary arguments as a basis for justifying individual or group values. Explaining how values arose was not the same as justifying them. Of those who took up the challenge of a more developed Darwinian ethics, the most well known was Leslie Stephen. His writings reflect better than any other author of the period the appeal as well as the inherent problems in Darwinian ethics.

Leslie Stephen

Middle-class intellectuals, particularly those who helped mold Victorian society in the latter decades of the nineteenth century by combining elements from the seemingly unlikely traditions of evangelicalism and rationalism, found in Darwin's writings a valuable key in their quest to formulate a new worldview. Central to this new worldview was a new system of ethics to replace the religious sanctions that for centuries had provided a guide for human conduct. Leslie Stephen, one of the editors of the volume of Clifford's essays, was representative of the liberal intellectuals who looked to science for a new foundation compatible with their utilitarian spirit and high moral tone. Stephen came from an evangelical family that belonged to the Clapham sect, which was famous for its efforts against the slave trade. Educated, genteel, but not wealthy, he belonged to what his biographer, Noel Annan, called the "intellectual aristocracy."[10] Ordained at Cambridge, Stephen broke with the church and became one of the leading agnostics of the age. In his politics, although he started off as a radical and supported reform, he feared mass party politics and held a great respect for authority. In later life he must have appeared reactionary to the younger generation of political reformers.[11]

Stephen, aside from being the unflattering model for Mr. Ramsey in To the Lighthouse, written by his daughter Virginia Woolf,[12] is remembered today largely for his editorship of the Dictionary of National Biography. In his day he was probably better known for his alpine climbing and promotion of mountaineering. But it was his work on ethics that he regarded as his most important contribution to the thought of his age. This work has not fared well, however. For most of the twentieth century philosophers ignored it, and even Annan disparagingly commented, "As a contribution to ethics, it is worthless."[13] Yet it represents the most developed elaboration in the nineteenth century of the Darwinian position on ethics, and it best reflects the internal problems its supporters faced.

As with many of his generation, Stephen was influenced by the utilitarian philosophy of John Stuart Mill, who stressed primacy of individual happiness as the major criterion for judging actions. Mill's formulation went far beyond the earlier, and cruder, utilitarians in expanding the notion of happiness from a simple pleasure-seeking and pain-avoiding of the greatest number view to one that stressed the quality of happiness and the importance of self-development in its social setting. Stephen, however, like Darwin, reacted against what he took to be the excessive individualism of that perspective. The Origin of Species initially suggested to him

10. See Noel Annan, Leslie Stephen: The Godless Victorian (New York: Random House, 1984): 5–7, and his "The Intellectual Aristocracy," in Studies in Social History: A Tribute to G. M. Trevelyan, ed. J. H. Plumb (London: Longmans, Green, 1955): 241–287. Also see Jeffrey Paul Von Arx, Progress and Pessimism: Religion, Politics, and History in Late Nineteenth Century Britain (Cambridge: Harvard University Press, 1985): 201.

11. For some of the political dimension of the Darwinist issue, see Greta Jones, Social Darwinism and English Thought (Atlantic Highlands, N.J.: Humanities Press, 1980).

12. An even darker side of this portrait is suggested in Louise DeSalvo, Virginia Woolf: The Impact of Childhood Sexual Abuse on Her Life and Work (New York: Ballantine Books, 1989).

13. Annan, Leslie Stephen, 286.

"a new armoury" to defend utilitarian doctrines, but in time he realized that such a move was inadequate and that a reconstruction of utilitarian ethics was necessary. He had carefully read the British moralists in preparation for his History of English Thought in the Eighteenth Century (1876), and before writing his ethics, he had benefited from Henry Sidgwick's careful analysis of utilitarian philosophy and critique of evolutionary ethics (see chap. 6). Stephen also had studied Spencer's works, and these helped convince him that what was called for was "a deeper change" of utilitarian ethics.[14]

Not that he followed Spencer's lead. As Stephen explained in the preface to his ethics, "Mr. Spencer has worked out an encyclopaedic system, of which his ethical system is the crown and completion. I, on the contrary, have started from the old ethical theories, and am trying to bring them into harmony with the scientific principles which I take for granted."[15] That is, he was attempting to extend utilitarianism with the new insights of Darwinian evolution.

It is worth pointing out that Stephen was not looking for a new guide for moral action. It was a rational justification for the principles of morality he accepted which he sought. Like Darwin and Clifford, he assumed that the notions of right and wrong were products of communities, not transcendental truths. Furthermore, he believed that in general terms these notions were not controversial. It was their justification that was the issue.

> No moralist denies that cruelty, falsehood, and intemperance are vicious, or that mercy, truth, and temperance are virtuous. Making every deduction from this apparent unanimity—allowing that similar names may be interpreted in very different senses; that the general outlines of a moral code may be the same whilst its spirit varies widely; and that the moral codes accepted at different times and places have been as different as has ever been seriously maintained—yet it remains true that there is an approximation to unity. The difference between different systems is chiefly in the details and special application of generally admitted principles. It is not such as we might anticipate from a radical opposition both of method and principle. But if we turn from the matter to the form of morality; if, instead of asking what actions are

---

14. Stephen discussed the background of his ethics in the preface to The Science of Ethics (London: Smith, Elder and Co., 1882).
15. Ibid., viii.

right or wrong, we ask, what is the essence of right and wrong? how do we know right from wrong? why should we seek the right and eschew the wrong?—we are presented with the most contradictory answers; we find ourselves at once in that region of perpetual antinomies, where controversy is everlasting, and opposite theories seem to be equally self-evident to different minds.[16]

Stephen looked to Darwin for the solution to this historical impasse and called for "a scientific treatment of moral questions." Breaking with a basic tenet of utilitarianism—the primacy of individual happiness—he insisted "that society is not a mere aggregate but an organic growth, that it forms a whole, the laws of whose growth can be studied apart from those of the individual atom."[17] This Darwinian shift from the individual to the group permitted him to approach the problem of ethics from a naturalistic perspective. The "scientific" resolution for the impasse in ethics, according to Stephen, was an evolutionary explanation of the nature of the moral sentiment. For him this entailed starting with "facts of observation." He wrote, "I have, that is, to consider the moral sentiments which have, as a historical fact, exercised an influence in the world, and to ask what part they play in the general process of evolution."[18]

Stephen was clear that he was not interested in "those moral principles which are or profess to be deduced from transcendental considerations or from pure logic independent of any particular fact."[19] These "ideal" moralities were far removed from the way people in fact acted and had led to confusion, misunderstanding, and conflict. What he was out to explore was "that set of rules which, as a matter of fact, is respected in a given society, and so far determines the ordinary approvals and disapprovals as to be an effective force in governing conduct."[20]

Stephen's ethics was a mixture of natural history, psychology, and sociology. Concepts like pain, pleasure, and sympathy were used to explain the determination of individual action, and these causes were embedded in a broader discussion of the variety of personality "types" found in a society. Just as the naturalists of his

16. Ibid., 1–2.
17. Ibid., 31.
18. Ibid., 36.
19. Ibid.
20. Ibid., 39.

day spoke of biological types as idealized morphological solutions to particular circumstances, he spoke of social types. Each type represented the most efficient adaptation relative to a particular situation. Stephen reasoned that since all actual organisms were imperfect, so, too, were all actual men. However, just as biological types improved in time, that is, the process of selection resulted in higher and more complex forms over time, so, too, did social types. Stephen was not advocating a straightforward march toward human perfection, for he stated that human excellence had many dimensions. Excellence as a poet and as an athlete, for example, although two types of excellence, were not necessarily linked and in fact might even be in opposition. Efficiency in one direction might imply deficiency in another.[21] The overall story, however, implied a successive net gain. The conduct of each social type was regulated by its purpose in life. What gave pleasure to one type did not necessarily give pleasure to all. Stephen explained that different character types might have different desires and therefore different motivations.

> There may, indeed probably there must be, properties common to all the typical forms, but they must be such as to be reconcilable with great individual variation. So, for example, a man may be a poet, a philosopher, a statesman, and so forth, and we may say that to each function there corresponds an appropriate type. Now it is conceivable that the highest excellence in different departments of conduct may imply consistent conditions. The greatest philosopher may also be the greatest athlete and the greatest poet. It is equally clear that there is no necessary connection. Brains of abnormal power may be associated with puny muscles. The sensibility of a poet, the preoccupation with abstract principles of a philosopher, may unfit either for business. . . . It seems, however, to be a highly general rule, that great excellence has, as it were, to be bought at a price, and that efficiency in one direction implies deficiency in others.[22]

But what determined character? Here Stephen moved beyond traditional British moral philosophy and advanced an evolutionary argument. He insisted that the only way to understand a person's character was by "referring to the conditions of existence. His character must be such as to fit him for the struggle of life."[23]

21. Ibid., 77.
22. Ibid., 81.
23. Ibid.

Human conduct, then, depended on social factors. One could not understand individuals without understanding the social forces that molded them. It was obvious to Stephen that "as every man is born and brought up as a member of this vast organization, his character is throughout moulded and determined by its peculiarities."[24] Using biological metaphors, such as the "social tissue," he related moral behavior to the social context. Stephen preferred to describe society as a tissue rather than as an organism. ᵢ ʟe meant to imply by the distinction that like a tissue, society was composed of individual units but that unlike an organism, it was not "capable of combining its efforts in order to bring about some common end."[25] Stephen noted that there were special combinations of the social tissue—what he called "organs"—such as nation-states, churches, and industrial organizations, but he downplayed their importance. Of greater significance was social competition, which existed in two major forms. One, a Malthusian struggle, occurred among the individuals of a race, for "given the faculties and character of the race, there is room for a certain quantity of population."[26] The second major type of competition was among races, and the result was group selection.

Stephen raised the interesting question, when should two races be considered of "identical social tissue"? He relied on his biological metaphor, not on physical characteristics, and proposed this criterion for determining the capacity of different races to blend with each other: "The organization of individuals should be regarded as identical when one individual could be transplanted into another race with perfect facility."[27] With typical Victorian blindness to racial assumptions, he referred to some "natural experiments," such as the rapidly blended "streams of population" in America drawn from different parts of Europe, in contrast to the "Chinese and the negro [who] remain side by side with the other populations instead of speedily losing their separate identity."[28]

No matter what persons composed it, society must have certain rules of conduct to maintain social relations. These customs of the race and habits of the individual were necessary for the life of the society and determined its identifying traits. Moral laws, then,

24. Ibid., 109.
25. Ibid., 126.
26. Ibid., 127.
27. Ibid., 131.
28. Ibid., 131.

were natural products of the evolution of the society and had developed over time. Stephen devoted several chapters to basic virtues like courage, temperance, truth, justice, and benevolence and illustrated their social significance. He concluded that "morality is a statement of the conditions of social welfare; and morality, as distinguished from prudence, refers to those conditions which imply a direct action upon the social union. In other words, morality is the sum of the preservative instincts of a society, and presumably of those which imply a desire for the good of the society itself."[29] But why should individuals work for the common good?

Stephen was thus brought to the critical ethical question of his generation, why should individuals behave for the benefit of society, especially if it is at a cost to themselves? What was the basis of moral sanction? Like earlier British moralists, Stephen had recourse to the concept of "sympathy" to help explain the origin and strength of the moral sentiment. It was through the ability to sympathize with fellow humans that we came to understand how others felt, how they regarded us, and how our actions affected them. Drawing on David Hume, he accepted the notion that conduct was based on feeling and that no strictly logical arguments could justify our moral intuitions. "Moral laws can only be deduced by some dexterous sleight of hand. 'Ought' and 'ought not' (as Hume somewhere says) are suddenly inserted in the place of 'is' and 'is not.'"[30] Similar to Mill and other nineteenth-century revisers of utilitarianism, who also relied on concepts like sympathy, Stephen enlarged the concept of individual happiness as the final basis for moral action. He did this by stating that although ultimately one did what made one happy, that happiness was conditioned by social sympathy and gave rise to "altruistic" actions. In place of the calculus of pleasure used by Jeremy Bentham and the early utilitarians, Stephen relied on the evolution of society to point to what was good for the health of society and to thereby provide the underpinning of moral behavior. Stephen was more

29. Ibid., 217.
30. Ibid., 313. David Hume discussed the is/ought distinction in his Treatise of Human Nature, reprint of original 1739 edition edited by L. A. Selby-Bigge (Oxford: Oxford University Press, 1960): Bk. III, Pt. 1, sec. 1. The interpretation of Hume's discussion is controversial. See W. D. Hudson, ed., The IS/OUGHT Question: A Collection of Papers on the Central Problems in Moral Philosophy (London: Macmillan, 1969).

Darwinian than Darwin in this regard, for Darwin had conceived of some of the noblest sentiments as nonadaptive.

The job of the evolutionary moralist, according to Stephen, was extremely complex.

> Every moral judgment, as I have argued, is an implicit (if not an explicit) approval of a certain type of character. It includes, therefore, an assertion that the highest type includes certain qualities of character, which, of course, imply corresponding modes of conduct. The highest type, again, must, according to our theory, be that which is on the whole best fitted for the conditions of social welfare. The problem is just as precise as the problem [of] which physical conformation is best adapted to satisfy the conditions of health, strength, and activity. The fact, so far as it is a fact, that we cannot obtain an accurate solution does not prove that there is no such solution to be found, but only that the solution requires longer observation and a more elaborate set of experiments before we can hit upon it. The experiment, in fact, is that which is being always carried on by the collective experience of the race; and though we have established beyond a possibility of doubt certain general principles which are the basis of the accepted moral code, there is still a considerable margin of uncertainty in details. Upon this assumption, the problem for the moralist is analogous to the problem for the artist; each is virtually trying to discover a certain type which had definite conditions to satisfy—briefly speaking, that of bodily vigour in one case and of social vitality in the other.[31]

Here we see that Stephen was not as open-ended in his thinking as Darwin had been. Darwinian evolution does not offer the hope of defining what physical constitution is best adapted to a particular environment. Adaptation is always relative. A plant or animal can improve its situation in different ways—exploiting different parts of the environment, avoiding competition, and so on. And, of course, the environment is always changing, so the situation is inherently dynamic, not static. Darwinian analogies to the animal or plant world, presumably, should avoid concepts like "highest type" or "best fitted."

Stephen also departed from Darwin's writings in another important way. If the moral sentiment was to be understood as a social phenomenon, what was the origin of the individual's moral sentiment? Darwin had proposed that a moral sentiment had de-

---

31. Stephen, The Science of Ethics, 339.

veloped from social instincts and had been refined by habit. Stephen seems to have fallen back onto the earlier reliance by writers like Mill and Bain on individual psychological development. Every form of association, according to Stephen, implied some training. In church, as a citizen, as an employee, one learned subordination, self-restraint, consideration, and other social discipline. Of greatest importance, of course, was the family. Reflecting a typical Victorian sentiment, Stephen wrote, "The true school of morality is the family, which represents a mode of association altogether closer, more intimate than any other, and in which there is not the same possibility of deviation from the moral code. The moral quality of every man is determined to a very great extent in his infancy."[32]

As humans, then, we were conditioned by our society and internalized its needs. This internalized prompting was what we called "conscience." Or, as Stephen put it, "The conscience is the utterance of the public spirit of the race, ordering us to obey the primary conditions of its welfare, and it acts not the less forcibly though we may not understand the source of its authority or the end at which it is aiming."[33]

Leslie Stephen believed that he had constructed an evolutionary ethics that surpassed utilitarianism by taking into account not only the individual dimension but also the social. His ethics, moreover, accounted for the evolution of moral codes. Human nature was not a constant but rather a variable that must be understood relative to the society and the level of its development. Like Darwin, he held that there had been an overall progress in human moral evolution.

Too versed in history and too disgusted with the politics of his day to conclude his ethics in a simplistic, triumphant tone, Stephen saw that duty would not lead always to happiness for the individual, nor would all men always naturally choose virtue, even when its evolutionary significance was evident. Science would not, therefore, improve behavior, nor was it necessary for science to provide a new sanction for morality.

> It is sometimes said that science cannot provide a new basis of morality; and this is urged as though it were an objection. I at least must

32. Ibid., 344.
33. Ibid., 350–351.

thoroughly accept the statement. What science proves, according to
me, is precisely that the only basis of morality is the old basis; it shows
that one and the same principle has always determined the development of morality, although it has been stated in different phraseology.
And, moreover, this principle is not the suggestion of any end distinct
from all others. The great forces which govern human conduct are the
same that they always have been and always will be. The dread of
hunger, thirst, and cold; the desire to gratify the passions; the love of
wife and child or friend; sympathy with the sufferings of our neigh-
bours; resentment of injury inflicted upon ourselves—these and such as
these are the great forces which govern mankind. When a moralist
tries to assign anything else as an ultimate motive, he is getting beyond
the world of realities.[34]

But if science, particularly evolutionary science, could provide
no new sanction, what could it do? Stephen's discussion, building
on Darwin's and Clifford's earlier writings, attempted to explain
the origin and development of a moral sense in man and to show
its social value. As to providing a guide to conduct or a philosophical justification for what we as individuals "ought" to do, it was
silent. He was clear that the path of virtue did not always lead to
happiness, and he was equally blunt in stating that some individuals would choose personal happiness over virtue when such a
choice had to be made. Moreover, there was no convincing rational argument that could be brought to bear on such an individual. The most that could be shown was that it was reasonable to
be moral. That did not upset Stephen. The last line of Science of
Ethics read, "But it is happy for the world that moral progress
has not to wait till an unimpeachable system of ethics has been
elaborated."[35]

Leslie Stephen, of course, personally knew right from wrong.
His evangelical background had inculcated a strong moral code,
and he was a self-conscious exemplar of how an agnostic could be
an upright and respectable citizen.[36] But many individuals were

34. Ibid., 461.
35. Ibid., 462.
36. The importance to him of demonstrating how an agnostic could be a moral
man was something that several writers on Stephen have noted. Keynes's biographer mentioned in his discussion of Stephen's death in 1904 that "although there
were many eminent agnostics at that epoch, none the less the maintenance of one's
position as such appears to have imposed a certain strain. . . . As a counterweight
to their unorthodoxy, the freethinkers seemed to need to build up defences; if one

less certain in their moral intuition. Worse, the period in which Stephen lived witnessed conflict that was a result of dramatic differences in value systems. What he took to be his moral strength could be regarded by others as his dogmatism. Similarly, his appeal to the universal conditions of man was so general as to allow the justification of contradictory actions. Stephen's relativism was of little help in solving concrete issues. As a philosopher commented later, "It has been demonstrated again and again that the Darwinian theory will lie down peacefully with almost any variety of ethical faith."[37]

Stephen's account of the development of moral virtues, similar to other Victorian attempts, was also more mythical than historical. But unlike traditional myths, his account lacked any emotional charge. After all, if one offended the Old Testament God, one might be pursued to the ends of the earth, one might face eternal damnation, and one's sins might be visited on future generations. By contrast, what was a crime against "nature"? Why should an individual sacrifice his pleasure for the abstract good of the community? Even if one's group suffered, it did so not in absolute terms but only relative to the benefit of another group of humans. Adaptation for the Darwinian was, after all, a relative condition. One's loss was another's gain.

This relativity was a difficulty for Stephen's ethics because he relied on the social utility of moral values to demonstrate their worth. An individual might not see the good of the group as an adequate sanction for proper behavior, but was the good of the group itself an adequate justification? As an evolutionary argument, Stephen was on solid ground when he insisted that groups would promote behavior that benefited survival and that groups with individuals who adhered to such standards would have a competitive edge. But was survival sufficient as a basis for ethics? As we will see in chapter 5, philosophers such as Henry Sidgwick argued that it was not. Stephen supplied no argument in its favor but rather, like Darwin, was satisfied to have shown the social origin of moral values.

---

was an agnostic, it was difficult to be just an ordinary simple person; one had to be especially high-minded and moral." See R. F. Harrod, The Life of John Maynard Keynes (New York: Harcourt, Brace, 1951): 172.

37. T. de Laguna, "Stages of the Discussion of Evolutionary Ethics," Philosophical Review 15 (1905): 583.

What, then, did Stephen's ethics accomplish? Ironically, like the Christianity it sought to replace, it provided a mythical account to justify a set of intuitions on proper moral behavior. It did not provide a sanction to validate those intuitions; it did not substantiate any particular code; nor did it validate accepting evolutionary considerations as a basis for ethics. By its generalness, it was little more than a spiritual sop to the already self-defined "virtuous man." As an elaboration of a Darwinian worldview, it provided a perspective from which to regard morality as a functional behavior. But as an ethics, it was neither more powerful than the utilitarianism it sought to extend nor more corroborated than the Christianity he dismissed.

Leslie Stephen's ethics, although the most developed Darwinian ethics written, was not the only major evolutionary ethics developed in the latter part of the nineteenth century. Herbert Spencer, whose writings were referred to by Stephen, formulated an evolutionary ethics parallel to Darwin's. The aim and thrust of Spencer's ethics differed significantly from Darwin's, but as we shall see, it suffered from many of the same criticisms.

3. Evolutionary Ethics

Herbert Spencer

Charles Darwin played a key role in the history of evolutionary ethics. It was his theory that convinced the general public that life had evolved. Furthermore, it was his emphasis on natural selection as the primary factor in evolution that survived as the central dogma of the theory of evolution. Since most versions of evolutionary ethics claimed to be, at a minimum, consistent with the theory of evolution, Darwin's perspective exerted—and has continued to exert—considerable influence on ethical formulations based on evolution. Darwin's discussion of the origin of the moral sentiment, moreover, was an invitation to consider ethics from an evolutionary point of view. It was secular and compatible with many middle-class notions. But the approach proved to be limited. Most serious was its failure to resolve two of the main issues in traditional moral philosophy: sanction for individual action, and justification for the foundations of ethics. Darwin, Clifford, and Stephen saw in evolution a secular explanation that supported their British, middle-class morality. They assumed that the long course of human history had resulted in moral progress and that the British gentleman was among the highest expressions of this moral development. The virtues of middle-class Britain were assumed to be the natural products of centuries of evolution from savage times. But however much they may have positioned themselves within the mainstream of moral opinion, they did not articulate a philosophy that adequately resolved central questions that moral philosophy was expected to address. Moreover, the position was ambiguous about the extent to which adaptation was the criterion of contemporary moral value. And if, as was the case for Darwin, some of the highest moral sentiments were not adaptive, then on what grounds could they be justified? Why should individuals, or society, protect animals and keep them from needless suffering?

Spencer's Ethics

Darwin and those who claimed to be applying his ideas of ethics were not the only ones approaching the subject from an evolutionary perspective. The first episode in the history of evolutionary ethics owed as much to Herbert Spencer as it did to Charles Darwin. Spencer's extensive publications, in fact, had a wider and more popular impact at the time. Alexander Bain wrote concerning Spencer's Data of Ethics (1879), "From this time forward, the Ethics of Evolution occupied a place in the standing controversy respecting the true foundation of an ethical system."[1] And in his review of that work, Bain claimed Spencer was the first to work out a new doctrine of ethics based on evolution.[2]

Spencer was more interested in ethics than was Darwin. Whereas for Darwin, man's moral sentiment had been an interesting problem in natural history, for Spencer, the origin and validity of morality was central to his entire philosophical system. Spencer stated in his Autobiography that his early writings began with examining a "politico-ethical" question, and although the issue led him to many related topics, he always returned to it, albeit in increasingly more advanced forms.[3] Spencer's career has been correctly characterized by historians as a quest for the formulation of a new set of absolute rules of conduct and a crusade for them.[4] These rules, based on man's moral sense, proscribed individual conduct and defined the limits of the state.

The desire for a complete understanding of the moral basis of society certainly led Spencer very far afield: to a study of the physical, psychological, and social evolution of man. Spencer started his writing career with a consideration of how man ought to live, and then, with breathtaking ambition, he set out a cosmic survey that not only adumbrated the historical evolution of the solar system, the earth, and the earth's inhabitants as well as the psychic, social, and political development of man but also attempted to uncover the underlying process that was responsible for that evolution, before returning to the topic of proper living.

1. Alexander Bain, Autobiography (London: Longmans, Green, 1904): 341.
2. Alexander Bain, "The Data of Ethics," Mind 4, no. 16 (1879): 569.
3. Herbert Spencer, An Autobiography (New York: D. Appleton, 1904), 2: 321–322.
4. David Wiltshire, The Social and Political Thought of Herbert Spencer (Oxford: Oxford University Press, 1978): 39.

Social Statics; or The Conditions Essential to Human Happiness Specified, and the First of Them Developed was Spencer's first book.[5] In it he laid out the central theme that provided him with a framework for his prolific outpouring in the next half century: the construction of a "strictly scientific morality."[6] His main interest was to establish the intellectual foundations of a moral society, and he viewed the subject of ethics from that perspective. Although sympathetic with much in the utilitarian tradition, Spencer explicitly rejected contemporary formulations that "the good" could be known by calculating what would bring the greatest happiness to the greatest number. Such calculations, he thought, were too shortsighted and psychologically unrealistic. Rather, he contended that "the moral law of society, like its other laws, originates in some attribute of the human being."[7] But what did we know of this attribute, man's moral sense? General introspection, he stated, could not provide a satisfactory understanding of it or the basis of morality, for humans were too variable and imperfect. What was necessary was to derive rationally a first principle to serve as a foundation.

His derivation was constructed by asserting a set of assumptions from which he drew a desired conclusion. The argument ran as follows.

> God wills man's happiness. Man's happiness can only be produced by the exercise of his faculties. Then God wills that he should exercise his faculties. But to exercise his faculties he must have liberty to do all that his faculties naturally impel him to do. Then God intends he should have that liberty. Therefore he has a right to that liberty.
>
> This however, is not the right of one but of all. All are endowed with faculties. All are bound to fulfill the Divine will by exercising them. All therefore must be free to do those things in which the exercise of them consists. That is, all must have rights to liberty of action.[8]

But a problem was obvious immediately. Unrestrained activity could lead to a clash of desires, where the fulfillment of one indi-

---

5. Social Statics (London: John Chapman, 1851) is often cited as having been published in 1850. It did, indeed, appear in December 1850 but carried an 1851 publication date. Spencer's writings went through several editions; however, there is no standard modern edition. Citations are from the authorized American edition (New York: D. Appleton, 1888), which was reissued numerous times.
6. Spencer, Social Statics, 13.
7. Ibid., 29.
8. Ibid., 93.

vidual's capabilities interfered with the exercise of another's. The first principle, therefore, as derived from his assumptions and modified to avoid contradiction, emerged as "Every man has freedom to do all that he wills, provided he infringes not the equal freedom of any other man."[9]

Spencer also provided a secondary derivation, based on human nature, of his first principle. In this argument he stated that "there exists in man what may be termed an instinct of personal rights."[10] This "selfish instinct, leading each man to assert and defend his own liberty of action,"[11] was balanced by a respect for the rights of others. Drawing on Adam Smith's Theory of Moral Sentiments (1759), Spencer noted that humans had the mental ability to imagine themselves in another's place. This "sympathy," according to Smith, was the root of our moral sense.[12] For Spencer, it was a key component, but more in the sense of a balancing sentiment that accounted for our sense of justice. Through a "sympathetic affection" of another's personal rights we come to respect the claims of other persons. On a more refined level, sympathy also was the root of our sense of beneficence, that is, the choice of actions that went beyond the recognition of the rights of others and attempted to increase (directly or indirectly) the happiness of others (suppressing a gratuitously sarcastic remark to a colleague or teenaged offspring, for example).

As a subeditor of The Economist during the writing of Social Statics, Spencer was sufficiently in touch with the day-to-day world to recognize that man's moral sentiment had not produced a just and beneficent society. But that did not disturb or surprise him. Man was imperfect. Morality was a guide to how man ought ideally to act. As man progressed his actions would be closer to the ideal. Spencer optimistically asserted that all of nature tended toward greater, ultimately complete, adaptability. This was true of the relationship of animals to their surrounding environment, and it was true in the mental and moral adaptation of man to the social environment. That man was not yet completely adapted merely reflected that "he yet partially retains the characteristics

9. Ibid., 121.
10. Ibid., 110.
11. Ibid.
12. See Adam Smith, The Theory of Moral Sentiments, in The Works of Adam Smith, vol. 1, reprint of 1811–1812 edition (London: T. Cadell) (Aalen: Otto Zeller, 1963). Also see T. D. Campbell, Adam Smith's Science of Morals (London: George Allen and Unwin, 1971).

that adapted him for an antecedent state. The respects in which he is not fitted to society are the respects in which he is fitted for his original predatory life. His primitive circumstances required that he should sacrifice the welfare of other beings to his own; his present circumstances require that he should not do so."[13] The future would see the emergence of the ideal man, "as certain as any conclusion in which we place the most implicit faith; for instance, that all men will die."[14] This point, that there was a gulf between what he later called "Absolute" and "Relative" ethics, served as a convenient strategy to propose a set of ideal principles that, if not practical, could still serve as a guide to action. "Right principles of action become practicable," he wrote, "only as man becomes perfect; or rather, to put the expressions in proper sequence—man becomes perfect, just in so far as he is able to obey them."[15]

Having established to his satisfaction a suitable foundation to serve as a guide to action, Spencer devoted the remainder of Social Statics to elaborating his social views. He discussed at length the rights of the individual: use of property, free speech, women's political rights, children's rights.[16] He also described in detail the limits of the state and in so doing expressed his distaste for the patronizing attitude of the middle class toward the working class as well as his strong objection to government encroachment on private liberties in the form of poor laws, public health proposals, and colonial policy.

For a number of years after publishing Social Statics, Spencer's writings focused on broader issues than how man ought to live. The vast system of philosophy that he constructed earned him the reputation of being the synthesizer of his generation's ideas. And in important ways he was. Although his knowledge of contemporary and past writers was unsystematic and although he made a point of not reading (but nonetheless discussing) authors with whom he disagreed, he was broadly informed on an enormous range of topics. His years in journalism provided him with extensive factual information in economics and politics. He made use of the large library of his club, the Athenaeum, and, perhaps most

13. Spencer, Social Statics, 77.
14. Ibid., 79.
15. Ibid., 51.
16. Spencer later altered some of his early liberal opinions on women. See Nancy Paxton, George Eliot and Herbert Spencer (Princeton: Princeton University Press, 1991): 7.

valuable, he frequented several brilliant, intellectual circles where he spent hours in conversation with friends and acquaintances with whom he conducted extensive correspondence. The list of his associates was impressive; it included John Stuart Mill, Thomas Henry Huxley, Charles Darwin, Joseph Dalton Hooker, George Henry Lewes, George Eliot, and John Tyndall. Many of his intellectual companions, such as Huxley, were personal friends. Spencer moved to John's Wood in 1858 in large part to be near Huxley with whom he took regular Sunday walks.

Spencer was also a member of the X-Club, a small dinner group that was a major political power in the world of British science.[17] At its monthly meetings Spencer was kept abreast of all that was intellectually fashionable in the world of science and culture. To a modern reader, Spencer's name looks somewhat out of place on the list of the nine members of the X-Club, as the other eight are eminent scientists.[18] All but Spencer were members of the Royal Society and held office in either the Royal Society or one of the other major scientific organizations. But Spencer was a valued member of the club, and his subsequent fall from fashion in the twentieth century should not cause us to underestimate his standing in the nineteenth.

Spencer's originality lay in his ability to combine in a readily readable (although not necessarily scintillating) form many of the concepts from others' writings. He did more than just compile knowledge. His achievement was a synthesis that appealed to those who were searching for a new scientific framework. As important as this achievement was, we should not lose sight of the basic goal of his work, which was to construct a new ethical vision for his contemporaries. It was no coincidence that his first book was a statement of an ethical position to serve as a guide for action and that his last, the culminating section of his Synthetic Philosophy, Principles of Ethics, was an elaboration of that statement.

The intervening volumes between Social Statics and Principles

17. See Roy MacLeod, "The X-Club: A Social Network of Science in Late-Victorian England," Notes and Records of the Royal Society of London 24 (1969): 305–322; J. Vernon Jensen, "The X-Club: Fraternity of Victorian Scientists," British Journal for the History of Science 5, no. 17 (1970): 63–72; and Ruth Barton, "'An Influential Set of Chaps': The X-Club and Royal Society Politics 1864–85," British Journal for the History of Science 23 (1990): 53–81.

18. The members were Edward Frankland, Thomas Archer Hirst, John Tyndall, Thomas Henry Huxley, George Busk, John Lubbock, Joseph Dalton Hooker, and William Spottiswoode.

of Ethics surveyed the broad terrain of the physical world, biology, man, mind, and society. First Principles, the opening volume of his sweeping series, A System of Synthetic Philosophy, established a general framework from which to organize not only ethics but knowledge in general. The key to such a unification, he believed, was the evolution principle, a principle inspired by his understanding of the ideas of the great German embryologist Karl Ernst von Baer. According to Spencer, evolution, in its most general sense, was a process that integrated matter and dissipated motion during which the matter "passes from an indefinite, incoherent homogeneity to a definite, coherent heterogeneity."[19] The direction of evolution was away from homogeneity, which he took to be an unstable equilibrium, to a more stable heterogeneity. It was his contention that this general principle, which drew on biological insights as well as current ideas in physics (thermodynamics), united all the separate branches of knowledge and could serve to synthesize them into a single natural process. The ultimate cause or nature of this principle was unknowable. Recognition of the phenomena of development in all areas of nature, however, sufficed as a justification for believing in evolution as universal.

The process of evolution led to greater diversity and individualization. Although First Principles stressed the evolution of the physical universe, Spencer was explicit that in addition to a cosmic evolution, there was a biological, a psychological, and a social evolution. Each of these evolutions was progressive in the sense that higher grades of evolution were more adaptive. Spencer's conception of evolution, in contrast to Darwin's, was not only broader but goal directed, self-corrective, and optimistic in that it assumed improvement and progress over time.

After setting out a general evolutionary philosophy, Spencer's Synthetic Philosophy covered biology, psychology, and sociology.[20]

19. Herbert Spencer, First Principles, 4th ed. (New York: D. Appleton, 1898): 407. The first edition was published in installments between 1860 and 1862.

20. Spencer published the first edition of his Principles of Psychology in 1855 before beginning the publication of his Synthetic Philosophy. His intent was to resolve the debate between J. S. Mill and William Whewell, that is, between empiricist and Kantian epistemologies. Spencer's resolution, which he believed was one of his most important contributions to philosophy, was founded on an evolutionary approach that traced the development of mind as successive adaptations to the environment. Mill's contention that all knowledge ultimately came from experience and Whewell's contrary view that knowledge was structured by a priori necessary

That is, after showing in his First Principles how the law of evolution applied in the physical world, he went on to demonstrate its unifying strength in the biological, mental, and social worlds. His method was basically similar in these volumes. He would start with a "Data of . . ." in which he would summarize the general phenomena of the subject; from there he would draw some inductive generalizations; and finally he would conclude with a synthesis from which the earlier facts could be deduced. At least, in principle. Even a cursory reading reveals that his phenomena were limited and carefully chosen, the inductions rather superficially generated, and the grand synthesis really not much more than a "causal" explanation, which was merely a translation into evolutionary language of the main generalizations.

Although Spencer's writings on cosmology, biology, psychology, and sociology were united, albeit in a labored manner, by his evolutionary philosophy, the driving goal—indeed, what underlay the entire system—was his vision of establishing a foundation for proper and just conduct. Central to his thought was the inviolability of the individual, a faith in the benefits of industrial society, and an abhorrence of war.

The importance Spencer attached to the moral dimension of his writings was reflected in the history of their publication. His prospectus of 1860, proposing a system of philosophy, consisted of five sections: first principles, principles of biology, principles of psychology, principles of sociology, and principles of morality.[21] Spencer had some concern whether his health would allow him to complete all the projected volumes, and in 1878, he interrupted work on the Principles of Sociology to tackle the concluding section of his system of philosophy. He published Data of Ethics in 1879 and in its preface explained his deviation from the originally proposed plan.

---

truths could both be accommodated by Spencer's synthesis. Spencer claimed that we were justified in holding a commonsense realism because it was the result of a long evolutionary process of successive adaptations whereby the mind adapted to its environment.

The First Principles of his Synthetic Philosophy established a broad evolutionary philosophy into which his psychology smoothly fit. Spencer expanded and incorporated the second edition of Principles of Psychology (1870–1872) into his Synthetic Philosophy.

21. The prospectus was reprinted in the preface to the first edition of the First Principles and continued to be reprinted in subsequent editions.

I have been led thus to deviate from the order originally set down, by the fear that persistence in conforming to it might result in leaving the final work of the series unexecuted. Hints, repeated of late years with increasing frequency and distinctness, have shown me that health may permanently fail, even if life does not end, before I reach the last part of the task I have marked out for myself. This last part of the task it is, to which I regard all the preceding parts as subsidiary. Written as far back as 1842, my first essay, consisting of letters on The Proper Sphere of Government, vaguely indicated what I conceived to be certain general principles of right and wrong in political conduct; and from that time onwards my ultimate purpose, lying behind all proximate purposes, has been that of finding for the principles of right and wrong in conduct at large, a scientific basis.[22]

Despite his fears, Spencer continued to live and write for quite some time. Data of Ethics was later expanded and incorporated into Principles of Ethics. Principles of Sociology was completed as well.

It was in his Data of Ethics and his more fully developed Principles of Ethics that Spencer set a moral foundation for his system of thought and a moral vision for his contemporaries. The basic approach that he took in this moral philosophy, however, had changed little from his statements in Social Statics. He still relied on the concept of sympathy, the distinction between Absolute and Relative ethics, and his optimistic vision of an ever-improving society. The main difference between his early and late moral writings, aside from the greater detail of the latter, was that he now grounded his ethics in an evolutionary philosophy, or perhaps more accurately, he now attempted to demonstrate how his moral philosophy was supported by a consistent and general systematic evolutionary philosophy. In Social Statics, Spencer claimed that God willed human happiness and that God was the source of man's moral sense. Although Spencer's writings continued to be amenable to a religious reading and were interpreted by many in that light, he substituted an evolutionary philosophy for his earlier deism and shifted to a fully naturalistic approach. Spencer pointed out the shift between his early and later ethics in the preface to the

22. The preface to Data of Ethics was reprinted in subsequent editions of the enlarged Principles of Ethics. See Principles of Ethics, 4th ed. (New York: D. Appleton, 1896), 1: xiii. The essay Spencer refers to was a set of letters originally published in 1842 in the newspaper Nonconformist in which he set out many of the ideas that he later published in his Social Statics.

fourth part of Principles of Ethics (1891): "What there was in my first book of supernaturalistic interpretation has disappeared, and the interpretation has become exclusively naturalistic—that is, evolutionary."[23] Spencer believed that there existed a pressing need for his naturalistic ethics. He wrote in the preface to Data of Ethics, "Now that moral injunctions are losing authority given their supposed sacred origin, the secularization of morals is becoming imperative."[24]

Spencer argued that conduct had evolved as part of the general evolutionary process.

> We have to enter on the consideration of moral phenomena as phenomena of evolution; being forced to do this by finding that they form a part of the aggregate of phenomena which evolution has wrought out. If the entire visible universe has been evolved—if the solar system as a whole, the earth as a part of it, the life in general which the earth bears, as well as that of each individual organism—if the mental phenomena displayed by all creatures, up to the highest, in common with the phenomena presented by aggregates of these highest—if one and all conform to the laws of evolution; then the necessary implication is that those phenomena of conduct in these highest creatures with which Morality is concerned, also conform.[25]

Spencer was able to substitute an evolutionary process for God as the foundation of his ethics rather easily, for he had been preparing the ground for years. In his biology and psychology, he had traced the adaptations of organisms to their environment and had shown that in organisms pain was correlative with injurious actions and pleasure was correlative with beneficial ones. There was, therefore, a natural evolutionary basis for seeking pleasure and avoiding pain.

In Principles of Ethics, he maintained his earlier postulate that happiness was the end goal of individuals by equating it with pleasure. Similarly, he retained his position that sympathy was responsible for part of the moral sentiment. But instead of viewing sympathy as a human faculty given to us by God, he explained it as the natural outcome of psychological and social evolution related to perceptions of pleasure and pain: "Pleasurable consciousness is aroused on witnessing pleasure" and "a painful consciousness is

23. Spencer, Principles of Ethics, 2: x.
24. Ibid., 1: xiv.
25. Ibid., 63.

aroused on witnessing pain."[26] Spencer had given a naturalistic origin of sympathy in his Principles of Psychology, tracing the sentiment back to "gregarious animals." Lack of intelligence limited the extent of sympathy in subhuman animals, but in man the potential was greater and had developed relative to social circumstances, first in a family setting and then in larger social groups.[27] In his later ethical writings Spencer continued to insist on his earlier distinction between ethical choices that individuals made affecting themselves but no others and those choices having an impact on one's fellowmen. With regard to the latter, he elaborated on those actions that concerned the rights of others, what he called justice, and those that were prompted by our sympathetic desire for others' happiness, what he called beneficence.

And, perhaps most important, Spencer continued to insist on the distinction between Absolute and Relative ethics. The difference was critical, for he believed that it allowed him to resolve the contradictions between ideal behavior and practical action, thereby providing him with a foundation for ethics that avoided the inconsistencies and inadequacies of other naturalistic ethical theories. It also took his ethical discussion beyond merely accounting for the origins of the moral sentiment to serving as a guide for proper conduct. Although for him, Absolute ethics applies only to "the completely adapted man in the completely evolved society," they nonetheless "serve as a standard for our guidance in solving, as well as we can, the problems of real conduct."[28] For example, in cases of what Spencer called "negative beneficence," that is, in moral situations where one's action had no direct bearing on another's welfare but nonetheless resulted in a negative influence on another's happiness—what we commonly call "unkind" actions. In such cases, Absolute ethics guided us "by enforcing the consideration that inflicting more pain than is necessitated by proper self-regard, or by desire for another's benefit, or by the maintenance of a general principle, is unwarranted."[29]

26. Ibid., 244.
27. Chapter 5 of the final section in the second volume of the Principles of Psychology is devoted to "Sociality and Sympathy." Spencer explored this topic in an article entitled "Morals and Moral Sentiments," first published in the Fortnightly Review (April 1871) and later reprinted in his Essays: Scientific, Political, and Speculative (1878).
28. Spencer, Principles of Ethics, 1: 275.
29. Ibid., 287.

Spencer had a significantly different approach to the moral sentiment than Darwin. Contrasting the two is somewhat complicated because they influenced one another.[30] Darwin, however, was primarily concerned to uncover the origin of the moral sense in order to show that the major distinguishing feature between man and the brutes could be encompassed by his biological theory. Spencer, although he engaged in the similar task of showing how the moral sentiment had come into being, was more interested in establishing the validity of this intuition. For by so doing, he was providing a scientific morality for his time and a justification for his social opinions.

Spencer and Darwin differed in other basic ways as well. Whereas Darwin thought that in animals some habits could gradually become instincts and that some acquired physical and mental traits could be transmitted, Spencer viewed life as a continuous adjustment of internal relations to external conditions. That is, Spencer was more Lamarckian than Darwin. Although Spencer denied that the future could be predicted in a specific manner, he was more teleological in outlook than Darwin, and he held that the action of natural selection was inadequate to explain evolution.[31] Rather, the organism, or in the case of man, the social group, actively adapted and the fittest survived.

Darwin's vision of evolution was less fixed than Spencer's. In Spencer's mind the outcome of social evolution was clear: a utopian, industrial society in which mutual aid replaced competition as the motive social force and in which the greatest individual freedom possible prevailed. Although Spencer's later writings reflected a gloomy vision of the immediate future, his overall vision never faltered. If it was true that a wave of militarism was spreading across Europe and threatened to undo many of the progressive advances of the nineteenth century and the specter of socialism, which he considered "biologically fatal" and "psychologically absurd,"[32] haunted Western nations, Spencer nonetheless was confi-

---

30. For an interesting discussion of the relationship, see John C. Greene's essay, "Darwinism as a World View," in his Science, Ideology, and World View, and Valerie A. Haines, "Spencer, Darwin, and the Question of Reciprocal Influence," Journal of the History of Biology 24, no. 3 (1991): 409–431.

31. Spencer, Principles of Psychology, 1: 615.

32. Herbert Spencer, Principles of Sociology, 3d ed. (New York: D. Appleton, 1898), 3: 582 (first published between 1876 and 1897).

dent about the ultimate outcome. In the conclusion to Principles
of Sociology he stated,

> If the process of evolution which, unceasing throughout past time, has
> brought life to its present height, continues throughout the future, as
> we cannot but anticipate, then, amid all the rhythmical changes in
> each society, amid all the lives and deaths of nations, amid all the
> supplantings of race by race, there will go on that adaptation of hu-
> man nature to the social state which began when savages first gath-
> ered together into hordes for mutual defence—an adaptation finally
> complete.[33]

Spencer did not think that a single higher society would come to
replace all existing ones. He tempered his earlier optimistic view
that the imperfect must disappear in time and that progress was
a social necessity.[34] Rather, in time different geographic regions
would harbor societies of varying degrees of sophistication and
the natural competition among societies would leave some inferior
"races" in less desirable locations. The great societies of the future,
however, would progress unhindered. Spencer reiterated his belief
expressed fifty years previously that "the ultimate man will be one
whose private requirements coincide with public ones. He will be
that manner of man who, in spontaneously fulfilling his own nature, incidentally performs the functions of a social unit; and yet is
only enabled so to fulfil his own nature by all others doing the
like."[35] In contrast, Darwin had little to say about the future of
man, nor did he predict any particular future political or economic
state. Although he held views about his own society, many, in fact,
that could be categorized as "Social Darwinism,"[36] he did not advocate a political or ethical vision in his publications.

Darwin and Spencer wrote quite differently about evolutionary
ethics, but often no distinction was made by writers who commented on the topic in the nineteenth century. Some analyses were
careful to distinguish Darwin's biology from his speculations on
the origin of the moral sense, and some pointed out the selectionist thrust of Darwin's writings from the more Lamarckian cast of

33. Ibid., 608.
34. Spencer, Social Statics, 83.
35. Spencer, Principles of Sociology, 3: 611.
36. See John C. Greene's essay, "Darwin as a Social Evolutionist," in his Science, Ideology, and World View.

Spencer's. But more often, Darwin and Spencer were lumped together, to the dissatisfaction of each. In time, however, the logical
differences were given more weight, and in the twentieth century
the two positions are generally clearly distinguished.[37] Their differences became more obvious partly due to the decline of the scientific status of the Lamarckism on which Spencer relied. With the
violent rejection of Lamarckism in the Anglo-American scientific
world of the twentieth century, Spencer's synthesis lost its scientific
underpinning, and this loss contributed to the obsolescence of his
writings. It would be misleading, however, to ascribe the decline of
Spencer's reputation solely to external factors like changes in biological theory. Of equal importance were the strong criticisms
aimed at his works for their lack of intellectual rigor and their
largely ad hoc arguments. His moral conclusions might have been
consistent with his premises, but they hardly followed deductively
from them as he claimed. We will examine more closely the philosophical arguments against his evolutionary ethics in chapter 5. It
will suffice here to note that Spencer's loss of popularity resulted
from both the devastating attacks on his work by philosophers
and the undermining of his assumptions by biologists.

Spencerian Evolutionary Ethics

Before his eclipse, Spencer's writings were viewed by many as
part of a set of texts that might serve as a foundation for a new
moral vision. Even some of those who rejected Spencer's overall
philosophy used portions of his writings to break away from traditional approaches to man and ethics. We have seen already that
Leslie Stephen, although he rejected Spencer's emphasis on the individual, was deeply impressed by his work. Alfred Barratt at Oxford was also inspired by Spencer. Although critical of Social Statics,
Barratt nonetheless borrowed heavily from Spencer for his Physical Ethics or the Science of Action: An Essay (1869). This stillborn
treatise attempted to establish the first principles of a new science

37. For example, Robert Richards's Darwin and the Emergence of Evolutionary Theories carefully distinguishes the two, as does Michael Ruse's Taking
Darwin Seriously: A Naturalistic Approach to Philosophy (Oxford: Basil Blackwell, 1986), and Derek Freeman's "The Evolutionary Theories of Charles Darwin
and Herbert Spencer," Current Anthropology 15, no. 3 (1974): 211–221.

of ethics on a psychological understanding of the origin of moral
sentiments and copiously quoted Spencer's psychology and physics.

The importance of a new foundation for ethics was widely felt
in the late nineteenth century, and a frequently encountered opinion was that without an acceptable solution to the "crisis in
morals," social chaos might ensue. There were many more substantial reasons to fear social unrest at the time, but the crisis
in morals was seen by intellectuals as a major source of instabil-
ity. William Henry Hudson, a writer and professor of English
who served for a time as Spencer's secretary, stated in his Introduction to the Philosophy of Herbert Spencer, "The supremacy of
the older, theologically-derived sanctions of conduct is breaking
down; and the danger, immediate and serious is, lest they should
be generally cast away as valueless and inefficient before any other
sanctions are established to take their place."[38]

Ironically, Spencer's two most well known popularizers both interpreted his writings in religious terms. In spite of Spencer's own
secularism, his ideas were portrayed in the guise of a revitalized
Christianity.[39] In Britain, Henry Drummond made his clarion call
the unity of science and religion. He held that "evolution and Christianity have the same Author, the same end, and the same spirit."[40]
Drummond was influenced by Spencer's ideas on the progress of
man, in particular, the view that the struggle for existence in human history will be replaced by altruism. Like the natural theologians of the late eighteenth and early nineteenth century, Drummond believed that moral lessons could be read in nature. He
went even further by asserting, "The position we have been led to
take up is not that the Spiritual Laws are analogous to the Natural
Laws, but that they are the same Laws. It is not a question of
analogy but of Identity."[41] Unlike the earlier and static natural

38. William Henry Hudson, An Introduction to the Philosophy of Herbert
Spencer (London: Chapman and Hall, 1897): 147.

39. James Moore, in his Post-Darwinian Controversies: A Study of the Protestant Struggle to Come to Terms with Darwin in Great Britain and America
1870–1900 (Cambridge: Cambridge University Press, 1979), examines a variety of
Christian evolutionary positions and provides an excellent background for understanding the popularizers of Spencer.

40. Henry Drummond, The Lowell Lectures on the Ascent of Man (London:
Hodder and Stoughton, 1894): 438. Also see George Allen Smith, The Life of Henry
Drummond (London: Hodder and Stoughton, 1899).

41. Henry Drummond, Natural Law in the Spiritual World (London: Hodder
and Stoughton, 1883): 11.

theology of William Paley, Drummond's natural history envisioned the progressive realization of a divine plan in the dynamic evolution of life on this planet.

Similarly, John Fiske, the leading popularizer of evolution in the United States, developed as the keystone of his philosophy what he took to be the religious aspects of Spencer's philosophy. Fiske contributed to his mentor's system by setting out a hypothesis on the origin of moral and social evolution of man based on the prolonged infancy of humans.[42] But Fiske interpreted Spencer in a manner wholly at odds with his secular philosophy, a tendency that deepened with time.[43] Although Fiske prided himself on what he considered his significant and original scientific contribution to the evolutionary origin of the moral sense, he emphasized repeatedly that evolution had to be seen in terms of God's immanence in the world and of man's destiny. Ethical intuition ultimately came from God and could not be treated as naturalistic. At a farewell banquet for Spencer on November 9, 1882, in New York City's Delmonico's, Fiske told his illustrious audience, "Mr. Spencer's work on the side of religion will be seen to be no less important than his work on the side of science, when once its religious implications shall have been fully and consistently unfolded."[44] An after-dinner speech was not the place to elaborate on religious implications, but he did indicate briefly that all religions accepted some divine power and accepted

> that men ought to do certain things, and ought to refrain from doing certain other things; and that the reason why some things are wrong to

42. Fiske first set out his hypothesis in 1873 in an article entitled "The Progress from Brute to Man," North American Review 117: 251–319. Later he incorporated it into his Cosmic Philosophy. See H. Burnell Pannill, The Religious Faith of John Fiske (Durham: Duke University Press, 1957); George R. Winston, John Fiske (New York: Twayne, 1972); Milton Berman, John Fiske: The Evolution of a Popularizer (Cambridge: Harvard University Press, 1961); and Jacob Lester, "John Fiske's Philosophy of Science: The Union of Science and Religion Through the Principle of Evolution," Ph.D. dissertation, Oregon State University, 1979.

43. Fiske's most mature statement on science and religion was in his Through Nature to God (Boston: Houghton Mifflin, 1899), where he stated, "Of all the implications of the doctrine of evolution with regard to Man, I believe the very deepest and strongest to be that which asserts the Everlasting Reality of Religion (191)."

44. Edward Livingston Youmans, ed., Herbert Spencer on the Americans and the Americans on Herbert Spencer. Being a Full Report of His Interview, and of the Proceedings of the Farewell Banquet of Nov. 11, 1882 (New York: D. Appleton, 1883): 52. The title of this work is misleading. Spencer left the United States on November 11; the banquet had taken place two days earlier.

do and other things are right to do is in some mysterious but very real way connected with the existence and nature of this divine Power, which reveals itself in every great and every tiny thing, without which not a star courses in its mighty orbit, and not a sparrow falls to the ground.[45]

Fiske went on to say,

> The doctrine of evolution asserts, as the widest and deepest truth which the study of Nature can disclose to us, that there exists a Power to which no limit in time or space is conceivable, and that all the phenomena of the universe, whether they be what we call material or what we call spiritual phenomena, are manifestations of this infinite and eternal Power. Now, this assertion, which Mr. Spencer has so elaborately set forth as a scientific truth—nay, as the ultimate truth of science, as the truth upon which the whole structure of human knowledge philosophically rests—this assertion is identical with the assertion of an eternal Power, not ourselves, that forms the speculative basis of all religions.[46]

Fiske was expanding on Spencer's concept that the ultimate, or first cause, of nature was unknowable. Spencer, in his First Principles, had suggested that in both religion and science an ultimate principle, the Unknowable, was a mystery into which the human mind could not penetrate. For him, this shared skeptical conclusion was a possible basis of reconciliation, and he continued to insist on it throughout his later writings.[47] Fiske, however, reified Spencer's Unknowable into a theistic God.

Spencer, who did not get to talk at length with Fiske after the dinner, wrote to Fiske shortly afterward.

> I wanted to say how successful and how important I thought was your presentation of the dual aspect, theological and ethical, of the Evo-

---

45. Ibid., 53.
46. Ibid., 55.
47. Spencer carried on a debate with Frederic Harrison, one of Britain's leading positivists, on the reconciliation of evolutionary philosophy and religion. Harrison claimed that Spencer's philosophy led to a negation of religion, and Spencer strongly reasserted his view that his philosophy did not negate religion as he understood it. The debate consisted of a series of articles in Nineteenth Century and the Pall Mall Gazette in the 1880s. The articles were collected and published by Gail Hamilton as The Insuppressible Book: A Controversy Between Herbert Spencer and Frederic Harrison (Boston: S. E. Cassino, 1885). On Harrison, see Martha S. Vogeler, Frederic Harrison: The Vocations of a Positivist (Oxford: Oxford University Press, 1984).

lution doctrine. It is above all things needful that the people should be impressed with the truth that the philosophy offered to them does not necessitate a divorce from their inherited conceptions concerning religion and morality, but merely a purification and exaltation of them. It was a great point to enunciate this view on an occasion ensuring wide distribution through the press; and if Youmans effects, as he hopes through the medium of a pamphlet reporting the proceedings, a still wider distribution, much will be gained for the cause.[48]

Spencer appreciated Fiske's efforts to bring his ideas to the general American public. And the way in which Fiske claimed that Spencer's ideas revitalized religion fit the time. But Fiske was moving in a very different direction. Spencer regarded religious sanction as an early stage in the evolution of man's ethical systems. Although he maintained a belief in the existence of the Unknowable, which could inspire a sense of awe, Spencer was emphatic in his opposition to any theology, ritual, or doctrine.[49] What was left for religion was the recognition of an unknowable cause behind appearances and a realization that earlier religious forms were the crude steps of man to adapt to an ever more complex social life. As for ethics, Spencer held that in time religion had to give way to a more satisfying rational and ideal basis for morality. Spencer could accept how someone like Fiske might smooth the path to a naturalistic interpretation for someone coming from a traditional background by explaining that the new philosophy was a refinement and extension of former orthodox beliefs. The reinterpretation of his work, however, to suggest the existence of an immanent presence of God in the world, the reality of an immortal soul, or value of worship was completely contrary to his ideas. He wrote to Edward Youmans in 1883 concerning a work by Drummond that was similar to Fiske's.

> I lately took up a book at the Athenaeum entitled Natural Law in the Spiritual World, by Henry Drummond. I found it to be in considerable measure an endeavour to press me into the support of a qualified the-

---

48. John Spencer Clark, The Life and Letters of John Fiske (Boston: Houghton Mifflin, 1917), 2: 264.

49. Bernard Lightman's The Origins of Agnosticism: Victorian Unbelief and the Limits of Knowledge (Baltimore: Johns Hopkins University Press, 1987) has a good discussion of Spencer's agnosticism. Spencer's solution to the long-standing conflict between science and religion through the recognition of a shared ultimate power resembles his equally superficial resolution of the philosophic debate between empiricism and Kantianism (see n. 20).

ology, by showing the harmony between certain views of mine and alleged spiritual laws. It is an interesting example of one of the transitional books which are at present very useful. It occurs to me that while the author proposes to press me into his service, we might advantageously press him into our service.[50]

In response to Fiske's Destiny of Man (1884), which asserted, "I believe in the immortality of the soul, not in the sense in which I accept the demonstrable truths of science, but as a supreme act of faith in the reasonableness of God's work,"[51] Spencer wrote diplomatically to the author, "You approach more nearly to a positive conclusion than I feel inclined to do."[52] Although Fiske was instrumental in making Spencer's ideas known in America, he fused those ideas with American transcendentalism and popularized a vision that included evolutionary ethics in a shallow and superficial manner. Like his historical writings, which glorified contemporary culture and projected it as the natural evolution of a cosmic plan, so, too, his picture of morality utilized the past to justify current ethical intuitions by reference to their alleged divine origins.

Fiske, like Drummond and other religious interpreters of Spencer, helped Spencer's evolutionary philosophy reach a wide audience. If we accept the notion that Spencer's goal in life was a scientific basis for proper conduct, then we have to see these popularizers in an ambivalent light. Although their writings might have spread Spencer's fame, they did so by negating his principal aim. Their works, in fact, point to Spencer's failure to provide a satisfactory naturalistic ethic. For by stressing a theistic reading of Spencer and attributing to God the origin of our ethical intuitions, they replaced the scientific basis of correct conduct with divine sanction. This substitution underscored the lack of emotional appeal in evolutionary ethics. To a cerebral type like Spencer, who had a moral vision of what society should be, evolutionary ethics was satisfactory, but for a wider audience, it evidently needed to be reinforced.

50. David Duncan, ed., Life and Letters of Herbert Spencer (New York: D. Appleton, 1908), 1: 309.
51. John Fiske, The Destiny of Man Viewed in the Light of His Origin (Boston: Houghton Mifflin, 1884): 116.
52. Clark, Life and Letters of John Fiske, 2: 322.

Whether or not Spencer's popularizers portrayed his ideas accurately, they did introduce him to many readers. Combined with his own extensive writings and their numerous editions, Spencer was a major figure in the history of evolutionary ethics. And the reaction against that position was often aimed at him.

Who were the critics? Those who rejected the rising tide of scientific rationalism viewed Spencer, in spite of his religious defenders, as a dangerous member of a new clerisy. Philosophers, as we shall see in chapters 5 and 6, found much to object to in his method and conclusions. Even among those whom we might expect to be his natural allies, the scientific community, there were grave misgivings. We shall look next at two critics from that community: Thomas Henry Huxley and Alfred Russel Wallace. Their criticisms epitomized two major objections to evolutionary ethics among scientists.

4. Darwinian Critics

Thomas Henry Huxley and Alfred Russel Wallace

Thomas Henry Huxley

In the middle of the last century, Thomas Henry Huxley, in what is now a well-known letter, wrote to his sister that he intended to leave "his mark somewhere and it shall be clear and distinct

> T.H.H., his mark." [1]

T. H. H. did leave his mark. Indeed, he left it all over the place, in almost canine fashion, during that century in which he so passionately acted. And his name, for better or for worse, has remained in the public mind.

Huxley is best known as Darwin's bulldog, but such an epithet scarcely captures his multifaceted and complex career.[2] For our story, such a characterization is ironic, for Huxley saw—and stated—many of the problems in approaching ethics from an evolutionary point of view. Huxley's critique, therefore, is of special interest, for in spite of any differences they may have had, he was the leading popularizer in Britain of Darwin's ideas.[3] Along with

---

1. Leonard Huxley, Life and Letters of Thomas Henry Huxley (New York: D. Appleton, 1900), 1: 69.

2. See William Irvine, Apes, Angels, and Victorians: Darwin, Huxley, and Evolution (New York: McGraw-Hill, 1955), and Cyril Bibby's two books, T. H. Huxley: Scientist, Humanist and Educator (London: Watts, 1959) and Scientist Extraordinary: The Life and Scientific Work of Thomas Henry Huxley, 1825–1895 (Oxford: Pergamon Press, 1972). An evaluation of Huxley's place in the world of science is found in Adrian Desmond, Archetypes and Ancestors: Palaeontology in Victorian London 1850–1875 (London: Blond and Briggs, 1982), and Mario DiGregorio, T. H. Huxley's Place in Natural Science (New Haven: Yale University Press, 1984).

3. How "Darwinian" Huxley was has been discussed many times. See, for example, Michael Bartholomew, "Huxley's Defence of Darwin," Annals of Science 32 (1975): 525–535, who describes the tension with Darwinian ideas in Huxley's

Darwin, he constituted the rootstock from which the basic orientation of contemporary evolutionary theory stemmed, that is, a changing world that could be understood in terms of nonteleological, mechanical, and material processes. Huxley, as previously discussed, also was a personal friend of Herbert Spencer and often advised him on biological issues. Huxley read, for example, the proofs of Principles of Biology to check that the zoological facts were correct. Their correspondence reflects a strong mutual attachment and concern. Huxley shared with Spencer, and Darwin, a broad common philosophical perspective that viewed the world primarily in scientific and material terms.

Huxley's writings portray an overall order in nature, which he found awe-inspiring. In his early works, he incorporated romantic ideas stemming from Thomas Carlyle and German philosophy when he described the cosmos, but by the 1860s, Huxley came to view nature in more empirical terms. Man was clearly part of this vision, and in three essays published under the title Man's Place in Nature (1863) he set out a view of man in the new intellectual landscape created by Charles Lyell and Darwin. Although cautiously done, the work was a polemic for accepting the idea of man's close affinity to other families in the primate order. "Thus, whatever system of organs he studied, the comparison of their modifications in the ape series leads to one and the same result—that the structural differences which separate Man from the Gorilla and the Chimpanzee are not so great as those which separate the Gorilla from the lower apes."[4]

Huxley's attempt to place man taxonomically with the higher apes shocked many Victorians and certainly annoyed the leading British comparative anatomist, Richard Owen, who believed otherwise. But Huxley was not nearly as radical as the German materialists of his day who wished to reduce man to a mere mechanical contrivance. Quite the contrary, Huxley wrote in this same work, "No one is more strongly convinced than I am of the vastness of the gulf between civilized man and the brutes; or is more certain that whether from them or not, he is assuredly not of them. No one is less disposed to think lightly of the present dignity, or de-

---

early thought, and Mario DiGregorio, T. H. Huxley's Place in Natural Science, who explores the ambiguity of even Huxley's more Darwinian latter years.

4. Thomas Henry Huxley, Man's Place in Nature, in his Collected Essays (London: Macmillan, 1893–1894), 7: 144.

spairingly of the future hopes, of the only consciously intelligent denizen of this world."[5]

Although Huxley may not have been as scandalous as Ludwig Büchner, the German materialist who was standard reading for German revolutionaries and Russian nihilists, his ideas certainly stood out in Britain. Huxley's article in the Fortnightly Review, "On the Physical Basis of Life" (1869), for example, caused a sensation and was heatedly discussed for over a decade.[6] But it was more than his startling pronouncements on the material basis of life, or on the relationship of man to the higher apes, that explain Huxley's celebrity. Rather, Huxley was noted as an articulate and outspoken advocate of the new professional scientific thinking that so revolutionized the Victorian world. By the 1870s, many held that science was the most powerful method of obtaining truth, and for some, it was the only valid method. The scientific cast of mind infiltrated in all areas, from manufacturing to theology. Although competing visions existed in opposition to this technical and analytical thrust, the secular, rational approach, called scientific naturalism, steadily gained ground through the century. Huxley, along with John Tyndall, Herbert Spencer, William Kingdon Clifford, and Francis Galton, provided the leadership in popularizing the value of a scientific approach to problems.[7] Since the mid-1860s, Huxley along with Spencer and Tyndall had been part of the influential and highly distinguished dinner group, the X-Club. This small but important clique promoted their vision by lobbying the members of the major scientific societies of Britain to adopt their professional standards and by combating conservative religious opposition to science.

Huxley spent much of his career bringing the analytical and empirical approach to the British public. He was concerned with institutional reform, with educational reform, and with popular

5. Ibid., 153.
6. Like Tyndall's equally famous Belfast Address to the British Association for the Advancement of Science, Huxley's materialist pronouncements have to be taken in the larger context of his other writings. Neither Tyndall nor Huxley were philosophical materialists. Huxley was very explicit in rejecting such views as unsound. He nonetheless stressed the pragmatic value of investigations of the material basis phenomena. On Tyndall, see Barton, "John Tyndall, Pantheist," 111–134.
7. See Frank Miller Turner, Between Science and Religion: The Reaction to Scientific Naturalism in Late Victorian England (New Haven: Yale University Press, 1974): 9.

education. Concerning the last of these, he stressed the value of a liberal education for the masses. For him, such an education was one that made wide use of science and that resulted in an intellect that "is a clear, cold, logic engine, with all its parts of equal strength, and in smooth working order; ready, like a steam engine, to be turned to any kind of work, and spin the gossamers as well as forge the anchors of the mind; whose mind is stored with a knowledge of the great and fundamental truths of Nature and of the laws of her operations."[8]

Although Huxley could be anthropomorphic in his descriptions of nature, it was the scientific method—empirical, analytical, experimental—that he contended was the main path to knowledge and the criterion for decision. Yet when it came to the matter of ethics, we see an odd tack by this proponent of rational empiricism.

Huxley did not follow Darwin in examining human morality as a problem in natural history. Not that he underrated the importance of Darwin's interest in explaining human behavior as part of a general explanation of life's evolution, but Huxley had additional concerns. He was deeply immersed in social issues, and much of his career was spent in promoting liberal reform. Although science was central to Huxley's agenda, both for its methodology and content, he viewed man's instincts from a different perspective. Human instincts may have had an adaptive value for earliest man, but human history, according to Huxley, was a story of man's efforts to go beyond his animal legacy.

Throughout his life Huxley grappled with the problem of reconciling his sense of justice with cosmic processes. In 1860, after his son, Noel, died, he wrote a long letter to the liberal clergyman Charles Kingsley outlining his views on immortality and cosmic justice. In it, he affirmed a belief that "Nature is juster than we."[9] There was an absolute justice in the system of things, and man had to surrender his will to "fact." Exactly what Huxley meant has been the subject of considerable discussion among historians. In more optimistic moods, he wrote of there being "happiness in ex-

8. Thomas Henry Huxley, "A Liberal Education; and Where to Find It" (1868), in Huxley, Collected Essays, 3: 86.
9. Huxley, Life and Letters, 1: 236.

cess of pain."[10] But as he grew older, Huxley seems to have regarded the forces of nature "just" in the sense of being universal and applied to all beings. Similarly, as he aged, his optimism increasingly was overcome by pessimism concerning the condition of man.[11] Late in life, in such a mood, he wrote his major work on evolution and ethics. Although other late essays adumbrate Huxley's views, his Romanes Lecture of 1893 and its publication the following year with an extended introduction (Prolegomena) as Evolution and Ethics, in which he argued that ethical and cosmic processes were antagonistic, must have surprised his followers. By this, he meant that in man's evolutionary history a state of art emerged that stood in conflict with the state of nature. Man perfected; man replaced struggle and change with stability and permanence. Huxley also argued that in addition to a contest between cosmic forces and the struggle of society for control over nature, there was an internal conflict, between man's instincts and his moral sentiments. Since man was from the animals, he retained instincts that evolution promoted: aggression and self-preservation. Civilization was an advance from that state. Through sympathy humans had come to empathize with one another, and through our higher intuitive moral sentiments we could transcend our more primitive instincts.

In his Prolegomena, Huxley underscored the opposition of human society and its mores to the biological realm by contrasting plants in a state of nature with plants under cultivation. He did this by way of an extended metaphor (fifteen pages!) showing the opposition of cosmic and horticultural processes. Nature produced endless change and "the competition of each with all."[12] By man's art, however, this cosmic process was challenged. In place of competition, there was an elimination of the struggle for existence. The horticultural process adjusted the conditions to the needs of the plants, restricted multiplication, and molded the environment to produce the desired result. One might object that the horticul-

10. Ibid.
11. James Paradis in his T. H. Huxley: Man's Place in Nature (Lincoln: University of Nebraska Press, 1978): 86, writes of Huxley as "a man of two visions, the one filled with hope and wonder, the other dominated by a sense of futility and doom."
12. Thomas Henry Huxley, Evolution and Ethics, in Huxley, Collected Essays, 9: 4. (Huxley's Evolution and Ethics has been reprinted by Princeton University Press [1989] with an excellent introduction by James Paradis.)

tural process could not be in opposition to the cosmic process since art was a product of man and man was a product of nature. Huxley answered, "If the conclusion that the two are antagonistic is logically absurd, I am sorry for logic, because, as we have seen, the fact is so."[13]

Nature, therefore, was no guide for ethics. Unlike comments made in his early years that man must submit to the overall order in nature, in these late writings, he clearly expressed the divide between man and nature. "Cosmic evolution may teach us how the good and the evil tendencies of man may have come about; but, in itself, it is incompetent to furnish any better reason why what we call good is preferable to what we call evil than we had before."[14]

In fact, in Evolution and Ethics, Huxley argued strongly against the extension of evolutionary ideas, especially natural selection, to the formulation of ethical standards. His motives may have had their origins outside of biological concerns, as will be discussed below, but this did not diminish the startling nature of his conclusions. One, after all, would have expected Darwin's bulldog and Spencer's peripatetic friend to propose a naturalistic ethic, partly because of its compatibility with a broad slice of British (principally Calvinist) theology and partly because of his commitment to a physically deterministic universe. If man had evolved as all other species had, then should not man be understood in natural terms? What other concepts could we use without endangering the entire edifice of Huxley's interpretation of science? The drift to naturalism, so much a part of Victorian culture, was especially evident in Huxley's writings, and it is difficult to imagine him accepting any agents other than purely physical ones. Yet Huxley emphatically rejected considering society a spiritual jungle as outlined and approved by the so-called Social Darwinists and castigated by Marx. Spencer, although he condemned the excesses of laissez-faire philosophy, particularly imperialism, optimistically believed that a noninterventionist state would ultimately lead to human fulfillment. Huxley was not so sure. He strongly supported governmental reform policies both at home and abroad. Trusting to the inevitable march of progress due to natural laws was not enough. Nor, for that matter, was nature a reliable guide for conduct. Huxley argued instead that man was locked in a battle with nature. "Let us

13. Ibid., 12.
14. Huxley, Evolution and Ethics, 80.

understand, once for all, that the ethical process of society depends, not on imitating the cosmic process, still less in running away from it, but in combating it."[15]

Why did Huxley depart so radically from his friend Herbert Spencer and from Darwinians, like Stephen and Clifford, with whom he shared so many other opinions? Although one of the main architects of the bulwarks of evolutionary theory, Huxley was not one to accept ideas uncritically or on authority. He disagreed with Darwinians on many points and was quite late in applying Darwinian notions to his own biology.[16] Nor was he hesitant to point out to Spencer what he believed was incorrect or misleading in the latter's writing. Of greater importance, Huxley had motives linked to his social and political views that brought him into disagreement with numerous friends. Thomas Henry Huxley spent much of his life in public. He was part of a small London elite that represented the new professional scientist, and he was deeply involved in striving to reform British society. Although fundamentally committed to the pursuit of science, he was equally interested in the theological, philosophical, and social questions of the day, many of which centered around what shape the new, more secular, increasingly middle-class British society would assume. The Metaphysical Society, which met in London between 1869 and 1880 to discuss "the many problems raised anew by the growing antagonism between religion and the critical spirit of science,"[17] was one of the numerous forums in which he expressed his ideas and which put him into contact with the leading thinkers in England who were concerned with the future of Britain. Both Leslie Stephen and William Kingdon Clifford were members, as were the more orthodox Archbishop Manning and William Gladstone. Many of Huxley's essays were published in the journal Nineteenth Century, owned and edited by one of the founding members of the Metaphysical Society, Sir James Knowles.

The new critical spirit that characterized the second half of the nineteenth century, especially the decade of the 1870s, undercut many traditional beliefs, and Victorians on both sides of the Atlantic were forced to reexamine the assumptions that supported

---

15. Ibid., 83.
16. See DiGregorio, T. H. Huxley's Place in Natural Science.
17. Alan Willard Brown, The Metaphysical Society: Victorian Minds in Crisis, 1869–1880 (New York: Columbia University Press, 1947): 10.

generally accepted moral ideas. For good reasons, then, historians write of the spiritual crisis of the 1870s. Huxley was among the active thinkers of this period who believed that reform of educational and scientific institutions was necessary. His liberal politics made him uncomfortable with the conservative views of those who sought to justify a laissez-faire, noninterventionist state policy by reference to natural selection in the realm of morality and social institutions. Herbert Spencer, in particular, had argued for a free reign of natural law in the human domain that was at odds with Huxley's reforming thrust. By the time he was writing Evolution and Ethics, Huxley was upset with Spencer's ideas (as well as with Spencer) and his followers.[18] He was equally disturbed by numerous proposed alternatives. During the last decades of the nineteenth century, social unrest had given the middle class grave worries from radical reformers representing the workingmen of Britain. The American, Henry George, for example, whose Progress and Poverty (1880) was widely read and discussed on both sides of the Atlantic, questioned private ownership of land.[19] William Booth's In Darkest England, and the Way Out (1890) dismissed the value of education and other social reforms in ameliorating the plight of the poor and proposed his Salvation Army program as the better solution. Huxley wrote a series of acerbic letters to the Times which questioned the tactics and assumptions of the program.[20]

Given Huxley's concern with political issues during the time he prepared his Evolution and Ethics, it is fair to say that he had

18. The story, as might well be imagined, is much more complicated. Spencer's and Huxley's relationship had been deteriorating since Huxley published his essay "The Struggle for Existence in Human Society" in 1888. Evolution and Ethics continued to argue against Spencer's noninterventionist view about extending the role of the state and against relying on the "cosmic process" to bring about improvement for man. Huxley also criticized the "fanatical individualism" of his day, to which Spencer took offense. However, Spencer, in rather typical fashion, did not see any contradiction to his overall theory of ethics; rather, he thought that Huxley was using material of his without acknowledgment. They were able with the help of some mutual friends, however, to patch up their personal differences. Spencer published a letter in the Athenaeum in 1893 pointing out the similarities between his ethical views and Huxley's. What he stressed was that they each thought natural selection inadequate to produce a moral sentiment. The letter was reprinted in his Various Fragments (New York: D. Appleton, 1898).

19. Huxley wrote to Knowles, "Did you ever read Henry George's book 'Progress and Poverty'? It is more damnder nonsense than poor Rousseau's blether. And to think of the popularity of the book!" Huxley, Life and Letters, 2: 261.

20. The letters are published in Huxley's Collected Essays, vol. 9.

ample political motivation for his antagonism to naturalistic ethics.[21] But he did not discuss those reasons, nor did he make an issue of them. Instead he focused on what he believed were the intellectual weaknesses of evolutionary ethics. Since his writings were widely read and his political motivation private, his published arguments were of greater direct historical significance for the discussion of evolutionary ethics.

What were his principal philosophical reservations? Although he could be strident and polemical, Huxley was acutely aware of the limits of human knowledge. He had read his Hume and Berkeley carefully and did not hold a simplistic, materialistic position. His famous "agnosticism" was symptomatic of a general skepticism that infused much of his work.[22] To be sure, there was what we would call today a "positivist streak" in his thought which claimed greater certainty for those ideas that could be observed or experimentally confirmed than for those that could not. But Huxley was critical of Comte's positivism and rejected the view that science held the key to understanding fully man and society. Following Hume, he also claimed that the leap from describing ethics from a natural history perspective to justifying those ethical precepts was not valid.

What mostly, however, set Huxley in later life against any evolutionary ethics was his belief that a conflict existed between man and nature. Numerous historians have explored Huxley's ambivalence toward nature and have pointed out the double conflict that concerned him, that is, the conflict between man and his environ-

21. In a well-known article by Michael Hefland, "T. H. Huxley's 'Evolution and Ethics': The Politics of Evolution and the Evolution of Politics," Victorian Studies 20, no. 2 (1977): 159–177, Huxley's Evolution and Ethics is portrayed as a "masterpiece of concealed debate" that attempted to deny the authority of the theory of evolution to writers like Spencer who championed the individual as well as to writers like Henry George and Alfred Russel Wallace, who espoused socialistic reform, but at the same time attempted to use the authority of evolution to bolster his own political views of domestic reform via a centralized, paternalistic state, along with a liberal imperialist foreign policy. Paradis correctly rejected this thesis in his introduction to the reprint of Evolution and Ethics: "Aside from giving more political unity and credibility to Huxley's opponents than is perhaps justified, [it] is too reductive for an individual of Huxley's experience and complexity." See James Paradis and George Williams, eds., Evolution and Ethics: T. H. Huxley's Evolution and Ethics with New Essays on Its Victorian and Sociobiological Context (Princeton: Princeton University Press, 1989).

22. See D. W. Docknell, "T. H. Huxley and the Meaning of 'Agnosticism,'" Theology 74, no. 616 (1971): 461–477. Also see Lightman, The Origins of Agnosticism.

ment, and the conflict within man between the instincts he had inherited and the cultural ideas that tempered them. More important, however, was the internal conflict between his vision of nature and his vision of man. Huxley was deeply committed to evolutionary theory and a physical view of life. Yet he also clung tenaciously to the idea of free will and constructed elaborate but unsuccessful attempts to account for human consciousness. As Alan Brown noted,

> Huxley, like most of the theologians and philosophers whom he criticized, was troubled by the very demon he sought to drive out: he as a scientist is so concerned with causality that he cannot conceive a thorough-going materialism without what he calls "necessarianism"; however, he finds it difficult to reconcile his own conviction of the power of individual volition with what he thinks a materialistic determinism demands. He is sure that every phenomenon has its efficient cause and admits the difficulty of proving any form of spontaneity, yet he is sure of the power of the will at least in part to determine or condition human phenomena.[23]

Like many of his seventeenth-century forebears, Huxley saw the dual nature of man, but unlike them, he had no deity to account for human mind and human morality. The issue had not become fully explicit in British science until the nineteenth century because previously supernaturalism was a part of the natural history tradition. The drift to naturalism, which characterized Huxley's thought and that of his peers, created the dilemma in biology. Whereas previously natural history and natural theology had been intimately united, by the 1870s that alliance was in shambles as a result of Lyell's destruction of Bible geology, Darwin's secularizing of the question of the origin of species, and the professionalization of science. Morality had been among the foundations of Victorian religious sentiment, and when the divorce of God from his creation occurred, morality was orphaned. For a scientist of Huxley's persuasion, the evidence for God was inconclusive and hence could not stand as the font of ethical systems. But neither could he see nature as a replacement. He was left with a wholly unsatisfactory evangelical legacy of childhood: that morality was intuitive emotion. In a work on Hume (1878), Huxley wrote, "In which ever way we look at the matter, morality is based on feeling, not on

23. Brown, The Metaphysical Society, 54.

reason; though reason alone is competent to trace out the effects of our actions and thereby dictate conduct. Justice is founded on the love of one's neighbour; and goodness is a kind of beauty. The moral law, like the laws of physical nature, rests in the long run upon instinctive intuitions, and is neither more nor less 'innate' and 'necessary' than they are."[24] The position was problematic. As James Turner noted, Huxley held moral principles as "universally binding on grounds that he himself would have derided as wishful thinking if used to maintain a belief about God."[25] For a man who had argued for "scientific naturalism," what could "goodness is a kind of beauty" mean? How could a purposeless universe give rise to a being with valid moral intuitions that were in opposition to cosmic processes that had produced him?[26] But Huxley was not alone in his inconsistency. Victorians, like Leslie Stephen, also confessed to knowing right from wrong independent of any rational argument. As mentioned earlier, the issue was never what man ought to do but rather how those intuitions could be rationally justified. Huxley saw that Christianity provided no foundation. Unlike many of his set, he also realized that the attempt to substitute an evolutionary foundation was just as invalid. Nor did utilitarian arguments, such as John Stuart Mill's, adequately address the internal conflict that Huxley perceived. Huxley's agnosticism, although unsatisfactory, was honest. He did not claim a rational basis for the values he held most firmly, and in that sense, he has been seen as foreshadowing modern writers like Faulkner and Camus.[27]

Huxley stood on the threshold of a tradition in evolutionary biology that began to explore the dilemma of man. He had the courage to state that evolution did not solve all the attendant problems, and he explored some of the consequences of that view. Whether or not he was motivated by cryptic, imperialist notions or not, his critical mind laid out basic issues for his fin de siècle

24. Huxley, Collected Essays, 6: 239.
25. James Turner, Without God, Without Creed: The Origins of Unbelief in America (Baltimore: Johns Hopkins University Press, 1985): 223. Also see A. O. J. Cockshut, The Unbelievers: English Agnostic Thought, 1840–1890 (New York: New York University Press, 1966), who describes the difficulty Victorians faced in attempting to find a substitute for Christian ethics.
26. See Greene, Science, Ideology, and World View, which discusses the paradox in Huxley's thought.
27. Cockshut, Unbelievers, 97–98.

audience as well as for us. Huxley's opposition demonstrates that one could be an ardent supporter of evolution, as well as scientific naturalism, and not embrace evolutionary ethics. There were also individuals who were solidly in the evolution camp but who rejected evolutionary ethics because they believed they had a firm alternative basis on which to ground their vision of man and his values. Alfred Russel Wallace exemplifies this dissenting group.[28]

Alfred Russel Wallace

Although Wallace independently formulated the concept of natural selection and continued to contribute actively to the development of evolutionary biology in important areas such as biogeography, he held different ideas concerning the nature of man than Darwin or Spencer. Unlike Huxley, Wallace's disagreement with the Darwinian picture was stated early on. In the 1860s, he advanced the opinion that natural selection had ceased to operate on the human body. As will be discussed below, Wallace modified this position twice in following years but continued to maintain that human evolution by natural selection was limited.

Wallace wrote in 1864 that with human mental and moral advance "man's physical character became fixed and almost immutable."[29] The aim of his paper was to undermine support for the position that the races of man were sufficiently different as to constitute different species, that is, the polygenesis theory of human races. The issue was highly charged, given the debates in the second half of the century over racism, slavery, and imperialism. If the races of man were different species, perhaps human moral obligations did not apply to them. Another topic Wallace addressed was why humans were so physically similar to living apes.

Wallace used a clever approach in his paper. He maintained that the evolution of man took place in two stages. During the

---

28. See Turner, Between Science and Religion, for an excellent discussion of a group of British intellectuals who characterized this position.

29. Alfred Russel Wallace, Contributions to the Theory of Natural Selection, 2d ed. (London: Macmillan, 1891): 176–177. Wallace's ideas were first presented in a talk he gave to the Anthropological Society of London, which appeared as "The Origin of Human Races and the Antiquity of Man Deduced from the Theory of 'Natural Selection,'" in the Journal of the Anthropological Society of London 2 (1864): clviii–clxx. The paper was reprinted with some alterations and additions in his Contributions to the Theory of Natural Selection (1870). Quotations are from the 1891 edition of Contributions.

first, natural selection operating on the earliest humans resulted in the production of different races. The second stage commenced with the development of man's mental and moral advancement, which was so powerful that his physical evolution ceased. This tack resolved both issues. Early in man's history physical differences between the bodies of living apes and man had ceased to increase because human physical traits were no longer influenced by natural selection. This similarity showed, moreover, that the conflict between the monogenesists and polygenesists was a pseudo-issue. Primitive man, according to Wallace, had a single origin in the distant past but had differentiated due to selective pressures acting on widely dispersed subpopulations to produce multiple races. The moral and mental evolution of man, however, marked such a discontinuity in nature that the physical characteristics separating races were insignificant. It was man's mental and moral faculties that constituted his true "humanness."

This was not to say that all humans were equally endowed. Early brutal races died out because of their lack of social harmony, and man's mental and moral faculties were still evolving. Wallace, citing Spencer's Social Statics, described a vision of a future, better adapted, and morally perfected humanity. Unlike Spencer, however, he still regarded as primary the force of natural selection. In time, Wallace believed that

> the higher—the more intellectual and moral—must displace the lower and more degraded races; and the power of "natural selection," still acting on his mental organisation, must ever lead to the more perfect adaptation of man's higher faculties to the conditions of surrounding nature, and to the exigencies of the social state. While his external form will probably ever remain unchanged, except in the development of that perfect beauty which results from a healthy and well organised body, refined and ennobled by the highest intellectual faculties and sympathetic emotions, his mental constitution may continue to advance and improve, till the world is again inhabited by a single nearly homogeneous race, no individual of which will be inferior to the noblest specimens of existing humanity.[30]

A few years later, Wallace claimed that the force of natural selection on man was limited, and historians have established that in large part he was motivated by a new interest in spiritualism.[31]

30. Ibid., 184–185.
31. See the excellent discussion in Malcolm Jay Kottler, "Alfred Russel Wal-

Spiritualism was a broad movement with thousands of adherents in the nineteenth century. George Bernard Shaw, whose mother held weekly séances, mockingly commented on the popularity of spiritualism in the preface to Heartbreak House, observing that the half century before the First World War was "addicted to table-rapping, materialization séances, clairvoyance, palmistry, crystal-gazing and the like to such an extent that it may be doubted whether ever before in the history of the world did soothsayers, astrologers, and unregistered therapeutic specialists of all sorts flourish as they did."[32] Spiritualism postulated the survival after death of the human spirit and the possible communication with these disembodied spirits. Many spiritualists were simple and unsophisticated; they attended séances for amusement or to search "for some incontrovertible reassurance of fundamental cosmic order and purpose, especially reassurance that life on earth was not the totality of human existence."[33] Other, more intellectual minds were searching for a surrogate religious faith appropriate in the modern world. A recent history of English spiritualism notes that "spiritualism appeared to solve that most agonizing of Victorian problems: how to synthesize modern scientific knowledge and time-honored religious traditions concerning man, God, and the universe."[34]

A year after his Anthropological Society of London paper, Wallace attended his first séance, and, impressed by what he saw, he began to read the spiritualist literature. He quickly was convinced of the reality of the phenomena as well as the validity of the spiritualist interpretations of them. For the rest of his life he remained a believer and became one of the main apologists in Britain for spiritualism.[35] His new beliefs altered the ideas he had about the

---

lace, the Origin of Man, and Spiritualism," Isis 65, no. 227 (1974): 145–192. Also see Janet Oppenheim, The Other World: Spiritualism and Psychical Research in England, 1850–1914 (Cambridge: Cambridge University Press, 1985).

32. George Bernard Shaw, Collected Plays with Their Prefaces (New York: Dodd, Mead, 1972), 5: 20–21.

33. Oppenheim, The Other World, 2.

34. Ibid., 59.

35. Wallace's credulity on the matter was notorious. On a trip to the United States, Wallace attended a séance where William James was also present. In his obituary notice, E. B. Poulton recounts a conversation with Josiah Royce who spoke with James after the séance. James had been quite suspicious and was surprised that Wallace had been so easily taken in. Royce quoted James, "It is a cu-

origin and nature of man and mind. In an unpublished letter of
1868 to the Pall Mall Gazette, he stated,

> I admire and appreciate the philosophical writings of Mr.
> Lewes, of Herbert Spencer and of John Stuart Mill, but I find in the philosophy
> of Spiritualism something that surpasses them all,—something that
> helps to bridge over a chasm whose border they cannot overpass,—
> something that throws clearer light on human history and on human
> nature than they can give me.[36]

In his published writings on the evolution of man up to his
famous book, Darwinism (1889), however, Wallace continued to
stress the empirical arguments for the limits of natural selection.
He enumerated various physical traits of man in 1870, either of
no use in adaptation or positively harmful, such as the general absence of hair covering man's body (especially on the back), so
striking in its departure from other terrestrial Mammalia. Such a
characteristic would be undoubtedly harmful to "savage man,"
and as evidence Wallace cited reports of voyagers who noted the
common habit of "savage races" to cover their backs with skins or
other protective coverings.[37] Similarly, he questioned the specialization and perfection of the hand and the presence of the organs
of speech.

His study of "savages" compared to anthropoid apes and civilized man led him to the more significant conclusion that "savages" and (by inference) primitive man had a brain size in excess
of any adaptive requirements: "A brain one-half larger than that
of the gorilla would, according to the evidence before us, fully
have sufficed for the limited mental development of the savage;
and we must therefore admit that the large brain he actually pos-
sesses could never have been solely developed by any of those laws

---

rious thing to see Wallace plunging head foremost into a flood which we Americans
only allow just to wet our feet." See E. B. Poulton, "Alfred Russel Wallace,
1823–1913," Proceedings of the Royal Society of London, Series B, 95 (1923–
1924): xxix.

36. Alfred Russel Wallace Papers, British Museum, add. mss. 46439, quoted in
Turner, Between Science and Religion, 88.

37. See "The Limits of Natural Selection as Applied to Man" in his Natural
Selection and Tropical Nature: Essays on Descriptive and Theoretical Biology
(London: Macmillan, 1891). This paper was first published in the 1870 edition but
contained ideas Wallace first suggested in 1869 in a review of new editions of two
books by Charles Lyell. See Kottler, "Alfred Russel Wallace," for a discussion of
this issue.

of evolution, whose essence is, that they lead to a degree of organization exactly proportionate to the wants of each species, never beyond those wants."[38] Man's brain size allowed for the potential mental sophistication that characterized civilized man. It was a developed organ that was without use for the "savage" whose languages, allegedly, lacked abstract ideas and who had no developed moral and aesthetic faculties. These latter would, indeed, be harmful to him "since they would to some extent interfere with the supremacy of those perceptive and animal faculties on which his very existence often depends, in the severe struggle he has to carry on against nature and his fellow-man."[39]

Wallace went even further. He argued that it was totally inconceivable that the capacity to form ideal conceptions of space and time, the intense artistic feeling for color, form, and composition, and the abstract notion of numbers, which makes possible arithmetic and geometry, could have "first developed, when they could have been of no possible use to man in his early stages of barbarism."[40] Contrary to what Darwin was about to publish in the Descent of Man, Wallace contended that "although the practice of benevolence, honesty, or truth may have been useful to the tribe possessing these virtues, that does not at all account for the peculiar sanctity attached to actions which each tribe considers right and more, as contrasted with the very different feeling with which they regard what is merely useful."[41]

What Wallace proposed was an innate moral sense, that "there is a feeling—a sense of right and wrong—in our nature, antecedent to and independent of experiences of utility."[42] For Wallace this moral sentiment was not accountable in Darwinian or Spencerian terms. Instead, he wrote, "The inference I would draw from this class of phenomena is, that a superior intelligence has guided the development of man in a definite direction, and for a special purpose, just as man guides the development of many animal and vegetable forms."[43]

38. Wallace, "The Limits of Natural Selection," Natural Selection, 193.
39. Ibid., 192.
40. Ibid., 199.
41. Ibid., 199.
42. Ibid., 201.
43. Ibid., 205. Oppenheim, Turner, and Kottler all point out that early in his life Wallace had been impressed by phrenology, which stressed that the mind possessed certain innate faculties, among them, the moral sense. On the development

Wallace still maintained that natural selection was the principal force in the evolution of nonhuman life. His overall conception of man's relationship to nature constituted a personal synthesis, and the asymmetrical juxtaposition of a chance process giving rise to plant and animal life coupled with a teleological process guiding human evolution did not trouble him. Good scientist that he was, however, he accepted criticisms of his position that natural selection could not account for certain of man's physical features. By 1889 when he published Darwinism, he modified his position accordingly. He stated,

> I fully accept Mr. Darwin's conclusion as to the essential identity of man's bodily structure with that of the higher mammalia, and his descent from some ancestral form common to man and the anthropoid apes. The evidence of such descent appears to me to be overwhelming and conclusive. Again, as to the cause and method of such descent and modification, we may admit, at all events provisionally, that the laws of variation and natural selection, acting through the struggle for existence and the continual need of more perfect adaptation to the physical and biological environments, may have brought about, first that perfection of bodily structure in which he is so far above all other animals, and in co-ordination with it the larger and more developed brain, by means of which he has been able to utilise that structure in the more and more complete subjection of the whole animal and vegetable kingdoms to his service.[44]

Natural selection, then, could account for man's physical state but not man's mental and moral faculties. Darwin's argument on the derivation of man's intelligence and moral sentiment, he wrote, "appears to me not to be supported by adequate evidence, and to be directly opposed to many well-ascertained facts."[45]

---

of British phrenology, which stressed the capacity of humans for improvement and leaned toward natural religion, see David De Giustino, Conquest of Mind: Phrenology and Victorian Social Thought (London: Croom Helm, 1975). Also see John D. Davies, Phrenology, Fad and Science: A Nineteenth-Century American Crusade (New Haven: Yale University Press, 1955; reprinted, Hamden: Archon Books, 1971).

As might be expected, Darwin was not pleased by Wallace's position. The most thorough Darwinian critique of Wallace's essay was published by the American Chauncey Wright. See his extensive review, "Contributions to the Theory of Natural Selection: A Series of Essays. By Alfred Russel Wallace, Author of 'The Malay Archipelago,' etc., etc.," North American Review 111 (1870): 282–311.

44. Wallace, Darwinism, 461.
45. Ibid.

There is a certain irony in Darwinism. As has been pointed out many times, Wallace was more Darwinian than Darwin in the sense of believing in the efficacy of natural selection to explain biological phenomena. But only up to man. Although he accepted the continuity that Darwin insisted on between the animal world and man, he maintained that man's moral and intellectual powers did not arise by natural selection. As one historian has dryly noted, "Darwinism presented the curious picture of fourteen chapters of neo-Darwinism followed by a last chapter of anti-Darwinism."[46]

Wallace admitted that the origin and nature of man's moral sense was a subject "too vast and complex" to be adequately treated in a short chapter. But, nonetheless, he went on to point out a few human intellectual characteristics—a mathematical faculty, musical and artistic faculty, plus a metaphysical and "peculiar faculty of wit and humour"—that were not associated with any adaptive value and that, moreover, were not uniformly present. This latter point Wallace took as significant because he believed that traits developed through natural selection were universal and relatively invariable among members of a species. The wide range of intellectual abilities suggested to him that natural selection was not their primary cause. His interpretation of these "facts" called for

> the existence in man of something which we may best refer to as being of a spiritual essence or nature, capable of progressive development under favourable conditions. . . . Thus we may perceive that the love of truth, the delight in beauty, the passion for justice, and the thrill of exultation with which we hear of any act of courageous self-sacrifice, are the workings within us of a higher nature which has not been developed by means of the struggle for material existence.[47]

Wallace was convinced that he had evidence of a spiritual world and that his interpretation of the origin of man's mental and moral faculties was not based on religious belief or speculation. He had seen and verified with his own eyes tables levitate, flowers and fruits appear on bare tables, and spirits manifest themselves. In his obituary notice on Wallace, E. B. Poulton reprinted a letter that contained the following remark: "I (think I) know that non-human intelligences exist—that there are minds disconnected from

46. Kottler, "Alfred Russel Wallace," 161.
47. Wallace, Darwinism, 474.

a physical brain—that there is, therefore, a spiritual world. This is not, for me, a belief merely, but knowledge founded on the long-continued observation of facts—and such knowledge must modify my views as to the origin and nature of human faculty."[48] His attempts to persuade scientific colleagues like Huxley, however, met with no success. Huxley, in response to an invitation to join a committee organized to investigate spiritual manifestations, wrote,

I regret that I am unable to accept the invitation of the Committee of the Dialectical Society to co-operate with a committee for the investigation of "Spiritualism"; and for two reasons. In the first place, I have not time for such an inquiry, which would involve much trouble and (unless it were unlike all inquiries of that kind I have known) much annoyance. In the second place, I take no interest in the subject. The only case of "Spiritualism" I have had the opportunity of examining into for myself, was as gross an imposture as ever came under my notice. But supposing the phenomena to be genuine—they do not interest me. If anybody would endow me with the faculty of listening to the chatter of old women and curates in the nearest cathedral town, I should decline the privilege, having better things to do. And if the folk in the spiritual world do not talk more wisely and sensibly than their friends report them to do, I put them in the same category. The only good that I can see in the demonstration of the truth of "Spiritualism" is to furnish an additional argument against suicide. Better live a crossing-sweeper than die and be made to talk twaddle by a "medium" hired at a guinea a seance.[49]

But Wallace had been a supporter of many unpopular ideas throughout his life. As a young man he had embraced mesmerism and phrenology and was strongly impressed by Robert Chambers's evolutionary hypothesis in the Vestiges of the Natural Creation. He also had been drawn early on to the utopian vision of Robert Owen and in spite of several decades' admiration for Spencer's writings, came to be a strong supporter of socialism.[50]

48. Poulton, "Alfred Russel Wallace," xxviii.
49. Huxley, Life and Letters, 1: 452.
50. See Alfred Russel Wallace, My Life: A Record of Events and Opinions, 2 vols. (London: Chapman and Hall, 1905), and James Marchant, ed., Alfred Russel Wallace: Letters and Reminiscences (London: Cassel, 1916).
An interesting discussion of Wallace's social views and how they relate to his science and spiritualism is in John R. Durant, "Scientific Naturalism and Social Reform in the Thought of Alfred Russel Wallace," British Journal for the History of Science 12 (1979): 31–58.

Wallace's support of spiritualism was more than just arguing for the reality of apparitions, clairvoyance, and table turning. The spiritualists' interpretation of these "facts" provided him with a perspective from which to unify his profound sense of social justice, his belief in the moral perfectibility of man, and his commitment to the rule of law in nature. The existence of higher intelligences that guided mankind explained what biologists had not been able to understand, and an enduring spiritual life after our physical existence provided an ethical sanction for this life.[51] Wallace accepted the spiritualist view that our state of happiness after death depended on our actions in this life but that in our future spiritual life moral progress continued, holding out the optimistic promise of a continued advance after the death of our physical body.[52]

In comparison with this prodigious Victorian edifice, evolutionary ethics was a pale prop for moral sanctions. Wallace had written to Spencer in 1879,

> I doubt if evolution alone, even as you have exhibited its action, can account for the development of the advance and enthusiastic altruism that not only exists now, but apparently has always existed among men. . . . If on this point I doubt, on another point I feel certain, and this is, not even your beautiful system of ethical science can act as a "controlling agency" or in any way "fill up the gap left by the disappearance of the code of supernatural ethics."[53]

Like Huxley, Wallace rejected the notion that the theory of evolution provided a basis for ethics. There was something in human nature that went beyond the quest for survival and was not to be explained in the same terms. Unlike Huxley, Wallace sought in spiritualism an explanation of this "something." The enormous popularity of spiritualism in the second half of the nineteenth century suggests that Wallace's concerns were widespread. The optimistic, progressive picture that spiritualism projected appealed to those, like Wallace, who sought justice and equity, if not in this world, then perhaps in the next.

Wallace's and Huxley's opinions on ethics show that all mem-

---

51. See his Miracles and Modern Spiritualism, rev. ed. (London: George Redway, 1896), for a full discussion of the "facts" of spiritualism as well as their interpretation.
52. Ibid.
53. Duncan, Life and Letters of Herbert Spencer, 1: 265.

bers of the Darwin camp were not supportive of evolutionary ethics. A commitment to the theory of evolution was not, therefore, a commitment to considering human culture in the same terms as biological phenomena. Their views also demonstrate that evolutionists who were attracted to Spencer's writings did not necessarily follow him to the conclusion of his ethicopolitical argument. The field of ethics was intimately connected with views of mankind, and there were many dimensions to the topic. Political, social, and intellectual factors influenced how people approached the subject of "right living." Although evolution was suggestive and various evolutionary options dealing with ethics were elaborated, it cannot be assumed that all supporters of the theory of evolution were drawn to, or supported, evolutionary ethics.

Having now considered the major versions of evolutionary ethics in the second half of the nineteenth century and two scientists who were critics, a more thorough examination of its contemporary evaluation is in order.

5. Early Reception and Evaluation of Evolutionary Ethics

1870–1890s

In spite of internal problems and some criticism by well-known authors like Huxley, evolutionary ethics proliferated and attained a wide hearing in the last quarter of the nineteenth century. Cora Williams, the author of a broad survey on the subject, characterized the situation with the following dramatic metaphor: "So many are the waters which now pour themselves into this common stream that the current threatens soon to become too deep and swift for any but the most expert swimmers."[1] In a burst of enthusiasm, stemming in part from her conviction that this new perspective on morality could serve as a justification for social reform, she added that evolutionary ethics "has unified and clarified the attempts made to discover a basis for moral principles and has rendered that foundation for the first time secure; it has cleared away, with one sweep, the rubbish of ancient superstition, made exact methods possible, and raised Ethics to the plane of a Science."[2]

Josiah Royce, of Harvard University, although he appreciated both the importance of the subject and the care with which Williams wrote on it, pointed out in his review of her book the problem of lumping together all the authors who considered ethics from an evolutionary perspective.

> The idea of treating in one place the evolutionary doctrines of ethics, and of writing their critical history, seems one eminently justified by the large place that evolution has played in recent discussion. Yet when one actually comes to consider the list of names brought together in the first part of this work, one begins to see how impossible it is to

---

1. Cora M. Williams, A Review of the Systems of Ethics Founded on the Theory of Evolution (London: Macmillan, 1893): 2.

2. Ibid., 515.

make, amongst recent writers, a distinct category of "evolutional" ethics thinkers, without hopeless perplexity.[3]

Royce was referring to the disparate character of the writings on ethics that argued from an evolutionary point of view. And his concern was legitimate. We have seen already how different Darwin's position was from Spencer's and have noted that subsequent authors were quite free in originating other versions of evolutionary ethics. A number of continental writers, who are outside the scope of this study because they belonged to separate philosophical traditions and had little impact in the Anglo-American world, constructed evolutionary positions on ethics as well. Such diversity of viewpoint, often proceeding from different intellectual or social goals, made the evaluation of the general approach difficult and contributed to the confusion that surrounded the subject.

The wide range of evolutionary ethics illustrates to what a large extent a new ethical theory was sought. A new rational foundation for morality was central to those who proposed a secular vision of man and society, and several of the keenest minds of the century struggled for decades with the issue. The writings of John Stuart Mill, James Martineau, Alexander Grant, Herbert Spencer, W. S. Lilly, and Alfred Barratt attest to the variety and depth, the synergism and originality, of ethical writings in the second half of the nineteenth century. Even for those who maintained that religion was the only legitimate basis for an ethical system, the need for a new approach to replace the old creeds was evident, and numerous theistic evolutionary systems of ethics, many allegedly based on Spencer, were proposed with the hope of revitalizing religious morality.

The Reception of Evolutionary Ethics

Given the multifaceted nature of the literature on ethics in the nineteenth century, it is hardly surprising that it is difficult to characterize the discussion on evolutionary ethics. Complicating the issue was the way in which a set of related issues—eugenics, "Social Darwinism," and various reform movements—were often woven into the arguments over evolutionary ethics. Consequently, all too

---

3. Josiah Royce, "Report on the Recent Literature of Ethics and Related Topics in America," *International Journal of Ethics* 3, no. 4 (1893): 535.

often today the nineteenth-century debate over evolutionary ethics is lumped together with associated but distinct topics and treated as merely one element in a general program of secularization that typified the new industrial culture of the Victorian period. Such a characterization, however, is hopelessly vague and obscures more than it reveals. Evolutionary ideas had a varied reception in different domains, and within each domain distinct concepts, such as "progress" or "selection," had different meanings. Although there is truth to the generalization that secularization characterized the Anglo-American culture of the second half of the nineteenth century and that evolutionary ideas were an important component of the trend, the family of ideas that were central to those intellectual controversies require close scrutiny to be understood, in the same way as an understanding of the Reformation demands a careful examination of doctrines and social context and is not adequately characterized by merely noting that during the sixteenth century a large portion of Europe became Protestant.

The diversity of the subject has not been the only impediment in characterizing the debate over evolutionary ethics. Another difficulty has been that modern academic disciplines, with their specialized categories and forums, were only just coming into existence. As a result, participants in nineteenth-century discussions of evolutionary ethics often appear to us to be at cross-purposes. They not only held different assumptions and goals but also functioned in different arenas. They therefore lacked a common vocabulary and a commonly agreed upon set of professional guidelines to channel the argument. In part, the difficulty in characterizing the "reception" of evolutionary ethics also reflects the different topography of the intellectual landscape of the last century. In England, for instance, a "revolution of the dons" occurred between 1860 and the 1880s during which the faculty of the ancient universities of Oxford and Cambridge began to take scholarship and teaching as serious duties of fellows. Before then, academic duties had not constituted a career but were a transitional period in a cleric's life from student to a church position.[4] An analogous revolution in American education started in the 1860s and by the

---

4. See A. J. Engel, From Clergyman to Don: The Rise of the Academic Profession in Nineteenth-Century Oxford (Oxford: Oxford University Press, 1983), and Sheldon Rothblatt, The Revolution of the Dons: Cambridge and Society in Victorian England (London: Faber and Faber, 1968).

end of the century, brought into existence the modern American university with its emphasis on public service, education, and research.[5]

Until almost this century, then, in both the United States and Britain, the major writers on ethics were outside the university and as with many sciences, were engaged in an "amateur" enterprise. Even by the end of the nineteenth century, the number of what we would call professional philosophers was small.[6]

This state of the discipline of philosophy has important implications for any current evaluation of nineteenth-century evolutionary ethics. Academic parochialism might prompt us today to refer the assessment of evolutionary ethics in the last century to our contemporary professional philosophers. Philosophers who specialize in the history of philosophy have not been shy about expressing opinions on subjects like nineteenth-century evolutionary ethics. Many such studies are historically sensitive and philosophically scrupulous in evaluating ideas using the philosophical standards of the time. But such a delegation of responsibility is highly problematic. As just indicated, the domain of philosophy was only beginning to be developed by the university, and therefore "the standards of the time" were standards held by a very small group. That that group would come to monopolize the field was not apparent, and therefore its standards cannot be facilely imposed on all the writings of the period. What standards can one use? Modern ones are clearly inappropriate, for our present highly specialized academics write primarily for other specialists. Modern university disciplines like philosophy have been dominated for long periods by specific perspectives and have focused on what professionals consider "critical problems." An unfortunate consequence of this trend is that most work done outside the current fashion is viewed with considerable suspicion, if not contempt. Major writers of the past often have been consigned unceremoniously to the gulags of the mind, only to be rehabilitated decades later as the precursors of contemporary insight. This is, of course, not a new phenomenon or unique

5. Laurence R. Veysey, The Emergence of the American University (Chicago: University of Chicago Press, 1965).
6. For an interesting discussion of the American philosophical community, see Daniel J. Wilson, Science, Community, and the Transformation of American Philosophy, 1860–1930 (Chicago: University of Chicago Press, 1990). The early issues of Mind, which began in 1876, have interesting survey articles on the state of philosophy in major British universities and in foreign countries.

to modern disciplines, for the present has always been selective in its judgment of the past. But the specialization of the academic world during the last hundred years does render attempts to evaluate past intellectual positions extremely difficult.

The only factor that simplifies the story is that there was considerable agreement in the late nineteenth century over what constituted the received morality (compared to what justified it). Owen Chadwick, who has written very sensitively of the moral dilemmas of the nineteenth century, tellingly observed in his discussion of the debate that was printed in Nineteenth Century on the topic "The Influence upon Morality of a Decline in Religious Belief" (1877) that "the burden sat heavier because nearly everyone, agnostic or not, assumed that the morality which they inherited was absolute and must be preserved, even though the creed linked with it might be dropped."[7] The issue, at least among the establishment, was not what we ought to do but why we ought to do it.

A critical review of the manifold Victorian reactions is likely to be the most instructive form of survey for evaluating the "reception" of evolutionary ethics in the late nineteenth century. For convenience, we can divide the reaction of those likely to be exposed to evolutionary ethics (i.e., the reading public) into three groups: the general public, the well-educated public, and the professional philosophers.

The general public in the nineteenth century considered the church the institution that properly dealt with moral topics. Not that there was much unanimity of opinion about what that meant. Different creeds, working-class apathy, and theological disputes made the Victorian church difficult to characterize.[8] Many of the more educated among the general public considered the morality of Scripture incompatible with their sense of moral justice. That is, they believed man could be improved and was not the depraved creature Christianity often made him out to be. These individuals wanted to reform religious teaching on ethics but with a few no-

---

7. Chadwick, Secularization, 231.
8. The literature on the subject is vast. A useful entrée is Chadwick, The Victorian Church. Also useful are Crowther, Church Embattled; P. T. Marsh, The Victorian Church in Decline: Archbishop Tait and the Church of England 1868–1882 (London: Routledge and Kegan Paul, 1969); and Anthony Symondson, ed., The Victorian Crisis of Faith (London: Society for Promoting Christian Knowledge, 1970).

table exceptions, like followers of Drummond and Fiske, rarely along evolutionary lines. The majority of the general public, however, never thought to question received religious authority. For them, morality was a set of commonsense guides to socially sanctioned action (e.g., goodness to others) and conformity to the strictures of specific creeds. An understanding of the systematic foundations of such behavior was of little concern.

Although the growing secularization of the century eroded the church's monopoly on morality, what replaced it was not a faith in rationality but an amalgam of ideas: typically a mix from early religious education plus the maxims of popular writers, like Samuel Smiles, on the norms of industrial society. Very unlikely (and unconscious) combinations of Comte, Paley, Spencer, and the Golden Rule were more widespread than adherence to the landmarks of Western ethics like Kant or Mill. As Chadwick dourly noted, "There is all the difference between a philosophical theory of ethics, which is always likely to be the property of a small group of specially equipped thinkers, and a system of ethics which has the potency to be accepted in society."[9] For most of the population, then, evolutionary ethics was not an issue with which they had to contend.

Only among the well-educated classes did there exist the potential for a serious discussion of evolutionary ethics, and in the learned literary journals and magazines one finds traces of it. The 1860s had been a time of considerable expansion in the publication of periodical literature, and for the educated reader, a number of high-quality reviews existed in the late nineteenth century.[10] When we look through those that dealt with philosophical and scientific subjects in addition to literary and/or religious ones, we encounter a modest number of articles on evolutionary ethics. There were, of course, reviews of all the major texts: Darwin, Stephen, Spencer, and so on. What is most striking to the modern reader about the reviews and articles is their lack of a common focus, or perhaps more accurate, the combined discussion of what today would be separate topics. Evolutionary ethics was discussed in ar-

---

9. Chadwick, Secularization, 230.

10. A good introduction to the relevant periodical literature is the classic work by Alvar Ellegård, "The Readership of the Periodical Press in Mid-Victorian Britain," Acta Universitatis Göthoburgensis 63, no. 3 (1957). Of enormous value is the 5-volume work by Walter Houghton et al., eds., The Wellesley Index to Victorian Periodicals 1824–1900 (Toronto: University of Toronto Press, 1966–1989).

ticles on positivism, Spencer's evolutionary philosophy, the relationship of science and religion, utilitarianism, "modern ethics," and the history of man. There was no sustained debate on evolutionary ethics per se, which did not reflect a lack of interest so much as a lack of specialization in the treatment of moral problems. The professionals had not carved up the field yet the way European nations were dividing other continents.

One should not conclude from the lack of specialization that nineteenth-century analyses were unsophisticated. To be sure, there were popular articles, but serious readers were quick to distinguish the level of analysis. One reviewer for the Westminster Review was moved (perhaps a little disingenuously, for the criticism could be leveled at him) to "enter our protest against this semi-popular style of discussing moral questions. Moral philosophy, or what is conceived as such, is just one of those subjects which is especially attractive to the general reader, but which the general reader is eminently unfitted to discuss."[11]

A review of the serious periodical literature reveals that Darwinian ethics was not generally distinguished from the broader views of Spencer[12] and that although biological evolution was very widely accepted, evolutionary ethics was not. Of the sixteen articles that touched on the subject of evolutionary ethics in the broadly representative Contemporary Review from its founding in 1866 to 1900, only a quarter were favorable. The majority strongly questioned the jump from describing an alleged history of the moral sentiment to proposing a moral code or its justification.

In contrast, supporters of Darwin and Spencer argued that morality was a problem that could be studied as a part of the natural history of man. Their perspective on ethics was conditioned by a general approach to understanding the natural world and was similar to the manner in which many students of history and ethnology examined the mores of "savages," ancient religions, or the culture of "primitive" man.[13] The validity of such studies was also

11. Richard Hutton, "The Natural History of Morals," Westminster Review, n.s., 36 (1869): 495.

12. A notable exception was Chauncey Wright. See, for example, his "German Darwinism," Nation 21, no. 532 (1875): 168–170.

13. For an interesting discussion, see Stocking, Victorian Anthropology. Stocking and other historians of anthropology have noted correctly the independent roots of the "classical evolutionary" position on the study of man and his history. Much

hotly contested, and even among those who were sympathetic to the point of view, the ethical significance of the results was highly debatable. Henry Calderwood (1830–1897), in an article published in 1877 in the Contemporary Review, clearly stated a widely held reservation concerning the relationship of natural science and ethics. He noted that biological science had progressed beyond mere classification and now not only sought laws regulating organic life but also proposed hypotheses concerning the origin and development of living forms culminating in the suggestion "that man is himself an illustration of this, being the last and crowning result in the history of evolution."[14] Calderwood was willing to concede to naturalists the domain of "the facts of animal existence" but stood firm in insisting that, although "the biologist claims to include all the facts of human life within the sweep of his own domain,"[15] he is quite ineffective in formulating a valid moral code, which instead is in the philosopher's province. Just as a philosopher "cannot argue from moral law to the determination of the facts of animal existence, it is just as plain that we cannot argue from a theory based only on facts observed in the lower forms of life to the determination of the facts of moral life."[16] Calderwood argued that attempts to formulate an evolutionary ethics were no more than reading our moral thoughts into nature. The theory of evolution "contains within it nothing but the barest scraps of material for an individual morality. It does not affect to be specially concerned with ethical questions, and does not profess to formulate an ethical theory. It is a biological theory, and that alone."[17] In his Evolution and Man's Place in Nature (1893), Calderwood agreed with Wallace that the living world had evolved but that biological evolution could not explain the emergence of the human mind or morality. He was particularly critical of Spencer's architectonic evolutionary synthesis, which he artfully disparaged. "In old Scot-

---

of what went for evolutionary interpretation was derived from nonbiological writers such as Mill. Many of these authors used a naturalistic, uniformitarian approach that stressed comparative methods and the interaction of human society with its environment.

14. Henry Calderwood, "Ethical Aspects of the Theory of Development," 31 (1877): 124.

15. Ibid.

16. Ibid.

17. Ibid., 131–132.

tish baronial architecture, certain overhanging turrets may be allowed, but these must not exceed the weight which the inner structure can endure and readily sustain."[18]

The Philosophical Community

Calderwood was a philosopher at the University of Edinburgh, and like other philosophers in the nineteenth century, he published articles for the public in the major intellectual journals. What did the other professional philosophers think? They were nearly unanimous in rejecting the validity of evolutionary ethics. The leading moral philosopher, Henry Sidgwick, argued that the entire approach was logically flawed, and subsequent philosophers who have criticized evolutionary ethics have for the most part merely provided variations of his analysis.

Sidgwick dominated British moral philosophy in the last quarter of the nineteenth century, and his writings set a new standard for the field. His work was historically of great importance, for it stood as a transition between classical utilitarianism and modern moral philosophy, and he therefore justly has been called the "first of the modern moralists."[19]

Sidgwick began his career at Trinity College and remained at the University of Cambridge for the rest of his life. He was sensitive to the tension that surrounded ethical debate, for he was sympathetic to the rival assumptions that separated writers on ethics. Many religious writers insisted that moral philosophy rested on intuitive maxims of divine origin, and even those religious thinkers who favored a more empirical ethics still relied ultimately on divine sanction. Opposed to these views were secular authors who contended that for ethics to be a rational enterprise it had to be in-

---

18. Henry Calderwood, Evolution and Man's Place in Nature, 2d ed. (London: Macmillan, 1896). I have quoted from the more interesting revised and expanded second edition. On Calderwood's context, see David N. Livingstone, Darwin's Forgotten Defenders: The Encounter Between Evangelical Theology and Evolutionary Thought (Edinburgh: Scottish Academic Press, 1987).

19. J. B. Schneewind, Sidgwick's Ethics and Victorian Moral Philosophy (Oxford: Oxford University Press, 1977): 122. Schneewind's study on Sidgwick is an excellent introduction to his work and context. Also see Arthur S. Sidgwick and Eleanor Mildred S. Sidgwick, eds., Henry Sidgwick: A Memoir (London: Macmillan, 1906).

dependent of all religious doctrines or assumptions. Most popular among the secular approaches was utilitarianism, which judged actions on their consequences—to produce or reduce pleasure. Sidgwick felt the tension between the secular and religious outlooks, and his hope that some reconciliation might be possible led him off into what today appear the strange byways of psychical research. He thought that through scientific investigation some empirical demonstration of the existence of the world of spirits might be possible. The existence of life after death, he believed, could provide a foundation for moral sanction. Sidgwick was the first president of the Society for Psychical Research (1882). However, unlike Wallace, he did not naively accept accounts of spiritualists, nor, although he attended many séances, did he ever witness any striking spiritualist phenomena.[20]

Sidgwick was drawn to religion but not into apologetics. As early as 1869 he resigned his fellowship at Trinity College because he felt that it would be hypocritical of him to appear to subscribe to the Anglican Church's thirty-nine articles, which was still a condition of fellowships at the University of Cambridge. Nevertheless, if not orthodox, Sidgwick was unwilling to abandon his religious leanings, for he saw no other possibility for the foundations of a rational system of conduct than a commitment to the existence of a God who guaranteed cosmic justice. John Maynard Keynes summed up Sidgwick's complex ambivalence to religion when he wrote to a friend, "He never did anything but wonder whether Christianity was true and prove that it wasn't and hope that it was."[21]

Sidgwick's major ethical work, The Methods of Ethics (1874), was an examination of what he considered the two major methods in contemporary moral philosophy: utilitarianism and intuitionism. In this classic text, Sidgwick defined the aim of ethics: "to

20. See the discussions in Oppenheim, The Other World, and Turner, Between Science and Religion. Oppenheim distinguishes spiritualism from psychical research. Spiritualists were those who believed in spirits and often sought contact with them. Psychical researchers sought to investigate claims of spiritualist phenomena and could approach the topic without preconceived opinions on the issue.

21. R. F. Harrod, The Life of John Maynard Keynes (New York: Harcourt, Brace, 1951): 116. Harrod quotes Keynes's letter to B. W. Swithinbank of March 27, 1906. This issue was more than just an interesting intellectual problem for Sidgwick. During the 1880s, in contemplating the lack of an adequate basis for ethical behavior, he experienced a severe emotional crisis. His struggle is documented in Sidgwick and Sidgwick, eds., Henry Sidgwick.

systematize and free from error the apparent cognitions that most men have of the rightness or reasonableness of conduct, whether the conduct be considered as right in itself, or as the means to some end commonly conceived as ultimately reasonable."[22] Most moral men believed, according to Sidgwick, that although their moral sense would generally be a good guide to action, there were "rules for determining right action in different departments of conduct: and that for these . . . it is possible to find a philosophical explanation, by which they may be deduced from a smaller number of fundamental principles."[23] Such an investigation was important. As Sidgwick reminded us, "For systematic direction of conduct, we require to know on what judgments we are to rely as ultimately valid."[24] The task for philosophers, then, was not in setting specific moral guideposts but rather in establishing a rational foundation for our actions. Sidgwick strongly believed that a system of ethics had to explain our commonsense moral judgment, and like the majority of his peers, he accepted the notion that our common moral inheritance was trustworthy.

Sidgwick attempted to synthesize the methods of ethics by arguing that although intuitionism did not provide ethical knowledge, an adequate utilitarianism must be grounded in a postulate of the moral order of the universe that could not be empirically demonstrated and could only be intuited.[25] Utilitarianism, accordingly, could justify accepted norms of behavior by reference to their utility in promoting happiness, but it could not account for basic altruism: why should we perform (or not perform) an action that will result in greater happiness for others if it is at our expense? To resolve this fundamental issue, Sidgwick believed that it was necessary to bring in an intuitive sense of some cosmic moral order. Sidgwick was under no illusion that he had provided a satisfactory synthesis for his age. His careful analysis of the different varieties and relationships of intuitionism and utilitarianism was highly valued by his contemporaries. But they, too, were not convinced that he had provided a meaningful synthesis. Indeed, across the channel Sidgwick's failure to provide an adequate ethical sys-

22. Henry Sidgwick, The Methods of Ethics, 7th ed. (London: Macmillan, 1907): 77.
23. Ibid., 102.
24. Ibid., 102–103.
25. Ibid., xx (preface to the 6th ed.).

tem was viewed by Marie Jean Guyau, the author of an important
and perceptive review of contemporary English moral philosophy,
as a sign of "crisis" in utilitarian philosophy.[26]
British opinion tended to be less harsh than Guyau's. Alexander
Bain wrote to Sidgwick praising the book highly and suggesting
that he should not despair of a solution. Bain suggested a recon-
sideration of the ideas that he had published on the psychological
foundations of ethics.[27] But to Sidgwick, a psychological inclina-
tion to do one's duty was not the same as a rational justification
of it.

Similarly, and of prime interest for us, evolutionary scenarios
were irrelevant for ethics according to Sidgwick. He indicated in
the preface to the Methods of Ethics that his book "avoided the
inquiry into the Origin of the Moral Faculty," for "if it be admitted
that we now have the faculty . . . it appears to me that the investigation of the historical antecedents of this cognition, and of its
relation to other elements of the mind, no more properly belongs
to Ethics than the corresponding questions as to the cognition of
Space belong to Geometry."[28]

Sidgwick was adamant about his position on evolutionary ethics, which followed from his definition of ethics. In response to
criticism that he had not treated evolutionary ethics sufficiently in
the Methods of Ethics, he published an article in Mind in which he
confessed that "all this is so evident, that what seems to need explanation is rather the fact that so much importance is commonly
attached to the question as to the 'origin of the moral faculty.'"[29]
He furthermore indicated his disdain for what he took to be the
uncritical attempts of authors to jump on the evolutionary band-
wagon. "Current philosophical notions characteristic of the most
recently accepted system or manner of thought in any age and
country are apt to exercise over men's minds an influence which is
often in inverse ratio to the clearness with which the notions them-

26. Marie Jean Guyau, La morale Anglaise contemporaine: Morale de l'utilité
et de l'évolution (Paris: Librairie Germer Ballière, 1879): 149.
27. Wren Library, Trinity College, University of Cambridge, add. ms. c.93.22,
Alexander Bain to Henry Sidgwick, 17 January 1875. Also see Bain's review of
Sidgwick, Alexander Bain, "Mr. Sidgwick's Methods of Ethics," Mind 1, no. 2
(1876): 179–197.
28. Sidgwick, Methods of Ethics, v–vi.
29. Henry Sidgwick, "The Theory of Evolution in Its Application to Practice,"
Mind 1, no. 1 (1876): 55.

selves are conceived, and the evidence for the philosophical doctrines implied in their acceptance is examined and estimated."[30] One can only imagine what he would have thought of our various postmodernist fashions.

His disdain, however, did not, nor had not, inhibited him from carefully examining the claims of evolutionary ethics. In the Methods of Ethics he stated what is usually considered the major objection, that "every attempt thus to derive 'what ought to be' from 'what is' palpably fails, the moment it is freed from fundamental confusions of thought."[31] He meant by this the garden variety "is/ought" distinctions, that is, either going from a description of a moral position to a belief in its validity, or going from a definition of the "natural" state of man (or society) to its use as a foundation for ethical principles. Sidgwick argued that the former position merely told us about customs and provided no justification of their validity. The latter, which continues to be hotly debated by contemporary philosophers, he dismissed as confused by noting that since any occurring impulse, tendency, or desire was "natural," how were we to distinguish those that were to guide us without employing some other ultimate (and unjustified) criterion, such as utility. Sidgwick's position, of course, rested on his demand that ethics have a rational foundation. Several contemporary philosophers have questioned such a demand. Some, like Alasdair MacIntyre, argue that an adequate ethics must rest on a broader philosophical foundation, that is, a metaphysical commitment. Others, like Richard Rorty, propose a return to pragmatism, while still others look to continental philosophy for either a hermeneutic option or to deconstruct the problem itself.[32] But it is unlikely that Sidgwick would have been very moved by such arguments, since they would be just other versions, for him, of the attempts to ground ethics in belief rather than reason.

To those authors, like Herbert Spencer, who sought to escape the criticisms against "naturalism" by reference to a cosmic process that was heading toward an ultimate state, the arrival of which we "ought" to anticipate and promote by appropriate actions, Sidgwick responded that an inevitable future no more deserved the opinion of what "ought to be" than "what is." In other words, by

---

30. Ibid., 52.
31. Sidgwick, Methods of Ethics, 81.
32. See chaps. 8 and 9 for discussions of contemporary positions.

appealing to some future state toward which we were inexorably moving—either directly or by fits and starts—we had not justified any claim of moral duty or established the value of any set of actions. What will be, will be; right, wrong, or indifferent.[33]

For Sidgwick, evolutionary ethics was not an ethical system or a method. It constituted merely an annoying popularized notion, not quite as vulgar as patent medicines but no more to be trusted. Sidgwick was outspoken in his opinion. His opposition was of considerable importance since he was a principal lens through which later generations focused on the topic. And considering that philosophy after Sidgwick's day came to be dominated by professionals, his influence at Cambridge cannot be underestimated.

Of equal importance in this respect was Thomas Hill Green at Oxford. Like Sidgwick, Green was among the first generation of professional philosophers who established the subject as an independent academic discipline. Hill was a personal friend of Sidgwick's, and as is obvious from their correspondence, each held the other in high esteem. They could not, however, have been farther apart philosophically. The author of a recent study of Green nicely captures the refreshing nature of their relationship when she notes that it "is a salutary example of what is missing most in philosophical dialogue today."[34]

Green founded a philosophical tradition at Oxford in conscious opposition to Sidgwick's. Unlike his Cambridge colleague, Green did not build his ethics on English utilitarianism but looked to Germany and the idealist tradition there, which stressed that the entire world was the product of "Absolute Creative Mind."[35] Un-

---

33. Sidgwick, Methods of Ethics, 83. Sidgwick elaborated on his critique of Spencer, especially of Spencer's later writings, in lectures that he gave at Cambridge. Although he had not prepared the lectures for publication, they were printed shortly after his death by E. E. Constance Jones, to whom they had been entrusted along with Sidgwick's other ethical papers. These detailed criticisms may have alerted Cambridge undergraduates to the imprecision and inconsistencies in Spencer's ethical writings, but they were less important to the general acceptance of Spencer's ideas than the broad rejection of Spencer's overall position found in the Methods of Ethics. The lectures appeared as Henry Sidgwick, Lectures on the Ethics of T. H. Green, Mr. Herbert Spencer, and J. Martineau (London: Macmillan, 1902).

34. Ann R. Cacoullos, Thomas Hill Green: Philosopher of Rights (New York: Twayne, 1974): 48.

35. This is not to imply that the Hegelian tradition was confined to Oxford. German idealism found supporters throughout the Anglo-American philosophical community. At Cambridge, F. H. Bradley and Bernard Bosanquet were major fig-

til World War I, Green's influence was paramount in Oxford philosophy, and if we hear little of him, and of idealism, today in the
Anglo-American world of philosophy, it is in large part due to the
violent, at times almost hysterical, reaction metaphysics like his
engendered after the war.[36] But before then, Green's Prolegomena
to Ethics (1883) was a major text that had to be reckoned with by
any serious student of moral philosophy. In it, Green argued for
the existence of a spiritual principle that manifested itself in man
and that could serve as a metaphysical foundation for ethics. Contrary to the then-popular associationist philosophy, which treated
knowledge as built up from sense perceptions, Green held that perceptions were related to one another by a self-distinguishing consciousness. According to his view, there could not be "perceptions"
without a "perceiving" agent. Our individual consciousness was
a part of an eternal consciousness that gradually realized itself
through humans. All knowledge, according to Green, presupposed
a simple system of relationships held together by an eternal consciousness. Although science, including associationist psychology,
could provide information on organic development, it was power-
less to illuminate the realm of knowledge, which was the task of
philosophy. Knowledge, including moral knowledge, could not be
understood in terms of natural history, especially not natural history as practiced by Charles Darwin; it could be understood only
in spiritual terms.

Green's metaphysics has often, and justly, been criticized as hopelessly vague, contradictory, and composed of an unstable combination of Aristotelian, Kantian, and Hegelian ideas. And when

---

ures whose philosophy informed an entire generation. In the United States, the
Journal of Speculative Philosophy, which was started in St. Louis under the editorship of William Harris in 1867, published translations of German philosophy and
original articles on romantic philosophy for over 20 years. Hegelians taught at major universities such as Johns Hopkins, Harvard, and California. See G. J. Warnock, English Philosophy Since 1900, 3d ed. (Oxford: Oxford University Press,
1978); William H. Goetzmann, ed., The American Hegelians: An Intellectual Episode in the History of Western America (New York: Alfred A. Knopf, 1973); John
Muirhead, The Platonic Tradition in Anglo-Saxon Philosophy: Studies in the
History of Idealism in England and America (London: George Allen and Unwin,
1931); and G. Watts Cunningham, The Idealistic Argument in Recent British and
American Philosophy (New York: Century Co., 1933).

36. For Green's reputation, see Melvin Richter, The Politics of Conscience:
T. H. Green and His Age (London: Weidenfeld and Nicolson, 1964). Also see Mrs.
Humphry Ward's novel, Robert Elsmere (1888), which not only reflects Green's
teaching but fondly portrays him under a fictional name at Oxford.

fashion shifted in philosophy, as it has every few decades, the new young Turks of the profession, such as Russell and Moore, did not have much difficulty demolishing Green's fragile intellectual construction.

But until after the war, Green's philosophy, in conjunction with Sidgwick's, defined much of moral philosophy. Neither was receptive to evolutionary ethics. Although Green believed that human history was a valuable adjunct to philosophy, for through it one could catch a glimpse of the realization of the divine plan, he did not interpret the history of man as did those who espoused evolutionary ethics. He agreed with Darwin and Spencer that there had been moral progress and that specific moral notions had changed in time. He also believed a conception of "the common good" characterized different historical periods and was, indeed, a precondition for the spiritual and political cohesion of its social groups. The validity of a moral code, however, was in its coincidence with the divine plan, not in its practical effects. Those practical effects, which for evolutionary ethics were the ultimate justification of the moral code, did not, according to Green, provide, in themselves, any moral obligation.

From the distance of a century, we can see that aspects of Green's moral philosophy were in a curious manner parallel with evolutionary ethics, rather like the way that natural theology and natural selection gave parallel explanations for "design" in nature. Green and most nineteenth-century proponents of evolutionary ethics recognized a progress in the history of moral beliefs, which they associated with the welfare of society. For Darwinians, of course, this progress was the outcome of a blind process, whereas for Green, and for teleological evolutionists like Spencer, there existed an overall process that was fundamental to an understanding of man's moral development. Green parted company with Spencer and similar teleological evolutionists not only in his metaphysical idealism but also in his answer to what for him was the basic question in ethics, how ought one to act? Or, what moral obligations were binding on an individual? The answer for Green resided in a free, self-willed conformity of the individual to his fulfillment of a part of the divine plan. However much he may have fudged his historical accounts to incorporate the values that he held on separate political and religious grounds, he attempted to justify them by an appeal to a spiritual principle. Green was emphatic that naturalistic philosophies, in contrast, could not carry moral obligation. In his Prolegomena to Ethics, he wrote, "It is obvious that to a being who is simply a result of natural forces an injunction to conform to their laws is unmeaning [meaningless]."[37] Indeed, according to Green, the attempt to account for morality by natural science results in "the elimination of ethics."[38]

Sidgwick and Green, the leading figures of professional philosophy, defined the task of ethics as a systematic inquiry into the foundations of the moral code. Their distinction between the moral code and its justification is one that is still basic (although occasionally challenged) in contemporary moral philosophy. Both Sidgwick and Green held that even if evolutionary ethics, Darwinian or Spencerian, explained the origins of a sentiment in man that could truly be called a moral sentiment, it did not furnish a justification for any of the moral prompting that followed from it. It was a mere black box that cranked out opinions. The attempt to justify those opinions with evolutionary arguments was no more than thinly disguised attempts to carry over other ultimate criteria, such as utility. But, as both Sidgwick and Green noted, such importations carried no moral obligation by themselves. They would have to stand the test of rational scrutiny. Green and Sidgwick each had his own idea of how an adequate ethics might be fashioned, and although neither has stood the test of time, ironically, their criticisms of evolutionary ethics, as the issue was then formulated, have.

The early reception of evolutionary ethics, such as it was, then, was not favorable. The general public did not welcome the initial spate of ethical writings inspired by an evolutionary perspective, nor did the well-educated classes or the professional philosophers. Sidgwick, in particular, provided a detailed analysis that demolished its philosophical standing. But Cora Williams, who was shortly to move from the east coast of the United States to Los Angeles and cease to expound on the subject (in favor of writing poetry), was not the only advocate for an evolutionary perspective on ethics. Evolutionary ethics, in spite of poor reviews, was to play to a large audience in the next few decades.

37. Thomas Hill Green, Prolegomena to Ethics, ed. A. C. Bradley (Oxford: Oxford University Press, 1883): 9.
38. Ibid., 10.

6. Evolutionary Baroque and Its Furies

1890–1920

The initial evolutionary writings on ethics by Darwin and Spencer and their supporters failed to elicit substantial positive reaction in the 1870s and 1880s. The first episode in the history of evolutionary ethics, as we shall see in this chapter, was far from over. Although informed opinion on the subject might be negative, the subject continued to attract attention. Its evaluation was complicated, however, by controversy over the theory of evolution itself. Naturalists, philosophers, intellectuals, clerics, and popular writers debated the validity of the theory of evolution in the decade following the publication of the Origin of Species. By the 1870s, the idea that animals and plants had changed over the course of time and that present-day living beings were the descendants of previous inhabitants of the earth was widely accepted in the Anglo-American world.[1] But as historians have noted, although Darwin won the war, he lost what to him was its most important campaign. Many scientists in the last quarter of the nineteenth century were skeptical of Darwin's main contribution to the theory: the idea that the principal force of evolutionary change was natural selection. Naturalists proposed alternative, complementary, or revised versions of Darwin's original formulation. At the turn of the century, there was so much theoretical disagreement that observers legitimately claimed that evolutionary theory was in a state of crisis.[2]

---

1. The story is, of course, quite complex. See Alvar Ellegård, "Darwin and the General Reader: The Reception of Darwin's Theory of Evolution in the British Periodical Press, 1859–1872," Acta Universitatis Göthoburgensis 64, no. 7 (1958); Edward Larson, Trial and Error: The American Controversy Over Creation and Evolution (Oxford: Oxford University Press, 1985); and Peter Bowler, "Scientific Attitudes to Darwinism in Britain and America," in David Kohn, ed., The Darwinian Heritage (Princeton: Princeton University Press, 1985): 641–681.

2. See Vernon Kellogg, Darwinism Today: A Discussion of Present-Day Scientific Criticism of the Darwinian Selection Theories, Together with a Brief Account of the Principal Other Proposed Auxiliary and Alternative Theories of Species-Forming (New York: Henry Holt, 1907), for a contemporary account of

Compounding the problem of the status of the theory of evolution was the profound realignment that the life sciences themselves were undergoing in the late nineteenth century. Traditional subjects that had defined the study of living beings were expanded by the use of new methods; new disciplines came into being; and areas that formerly had been separate studies were merged to create new ones. Taxonomy, which had been at the heart of eighteenth-century natural history, was given less attention by nineteenth-century naturalists, who shifted their focus to morphology and comparative anatomy.[3] More important, naturalists included physiology, which in the eighteenth and early nineteenth century was primarily associated with medicine, in their purview. By 1900, rigorous experimental methods associated with physiology were being used by researchers to shed light on problems of an expanded biological science. New problems in the investigation of inheritance, development, and psychology emerged and redirected researchers into areas that often were remote from any immediate evolutionary or medical significance but that in a few decades would lead back to the very sources from which they originated. At the turn of the century, however, "biology" seemed to be something new and different.[4] Part of the "crisis" of evolutionary theory was that the seemingly intractable problems regarding its foundations were not resolvable by experiments.[5]

Given the general situation at the time, it is not surprising that

---

the issue. Peter Bowler has discussed the topic with an eye toward some of the Lamarckians of the end of the century. See Peter Bowler, The Eclipse of Darwinism: Anti-Darwinian Evolution Theories in the Decades around 1900 (Baltimore: Johns Hopkins University Press, 1983).

3. See Paul Lawrence Farber, "The Transformation of Natural History in the Nineteenth Century," Journal of the History of Biology 15, no. 1 (1982): 145–152.

4. See Garland Allen, Life Science in the Twentieth Century (New York: John Wiley and Sons, 1975), who describes the changes as a revolution. A tempering of that claim appears in the "Special Section on American Morphology at the Turn of the Century," in the Journal of the History of Biology 14, no. 1 (1981). Since then there has been an explosion of high-quality studies on the history of biology at the turn of the century. A rich sampling is to be found in Ronald Rainger, Keith Benson, and Jane Maienschein, eds., The American Development of Biology (Philadelphia: University of Pennsylvania Press, 1988), and Keith Benson, Jane Maienschein, and Ronald Rainger, eds., The Expansion of American Biology (New Brunswick: Rutgers University Press, 1991).

5. The excitement over the rediscovery of Mendel's laws arose partly from the hope that experimental studies in genetics would permit evolution to become an experimental science. William Castle, for example, devised genetic experiments to elucidate the force of natural selection. See H. Terry Taylor, "William Ernest Cas-

the younger generation of evolutionary biologists did not actively engage in discussions on the ethical implications of evolution. The theory itself was problematic; the issue of ethical naturalism had ceased to be novel; those formally responsible for an educated opinion—the philosophers—had clearly rejected the position as illegitimate; and in an age when specialization was coming to be a criterion for legitimacy, philosophical speculation increasingly was considered an inappropriate activity for the "modern" scientist. What one finds in the writings of biologists concerning the topic is either a benign neglect or brief, highly general discussions. William Keith Brooks, the leading American Darwinist at the turn of the century, for example, in his philosophic Foundations of Zoology (1899), suggested that our moral sense had a natural history. He did not speculate, however, on that history but merely noted that we had nothing to fear from recognizing a natural origin of ethical beliefs.

Ironically, while the biologists between 1870 and the beginning of the twentieth century were uncomfortable with the theory, evolution as a general concept had become popular with the public. Writers invoked evolution as an explanatory idea to account for everything from the composition of biological communities to the alleged lack of intelligence among the new immigrant populations pouring into the United States from southern and eastern Europe. For numerous fashionable writers, "evolution" had the same explanatory power as the "Creator" in the previous century. This popular development resulted from the social changes of the Industrial Revolution and the consequent intellectual shifts that attended it as well as philosophical trends going back to the early part of the nineteenth century. And in spite of devastating critiques by professional philosophers, sweeping philosophical systems of writers like Spencer's were read widely and taken seriously by the public on both sides of the Atlantic. These systems were all prem-

---

tle, American Geneticist: A Case-Study in the Impact of the Mendelian Research Program," M.A. thesis, Oregon State University, 1973.

The significance of genetics for evolution, as has often been pointed out, was not clear. But the experimental nature of the study gave it an importance that has not diminished. See William Provine, The Origins of Theoretical Population Genetics (Chicago: University of Chicago Press, 1971), and Ernst Mayr and William Provine, eds., The Evolutionary Synthesis: Perspectives on the Unification of Biology (Cambridge: Harvard University Press, 1980).

ised on an evolutionary perspective and included an acceptance of biological evolution, albeit rarely of a Darwinian sort. The authors usually sidestepped technical biological issues and focused instead on the broader cosmological features of an evolutionary perspective, or on a speculative cultural anthropology that seemed to be ripe with implications for political and social thought.

It would be a mistake, therefore, to construe the lack of enthusiasm accorded evolutionary ethics by serious writers to be evidence that evolutionary ethics had not survived to the end of the 1890s. Although philosophers might write that "evolutionary ethics, as a peculiar variety or school, has almost ceased to exist,"[6] there were still writers who saw not only in the Spencerian but also in the Darwinian picture of nature a valuable starting point for social and ethical explanation. Like Leslie Stephen in the 1880s, several thinkers around the turn of the century attempted to construct or justify their moral intuitions by reference to biological evolution. In this chapter we will consider three examples of authors whose books were representative of how evolution was used to discuss social and ethical issues: Benjamin Kidd, who started from a Darwinian perspective but found evolution wanting as a basis for morality; Woods Hutchinson, who stayed closer to the Darwinian spirit and elaborated one of the most provocative positions at the turn of the century; and Samuel Alexander, who cobbled his own original Darwinian ethics. Two critics will then be considered. An examination of G. E. Moore's analysis of evolutionary ethics will reflect the British philosophical community's continued rejection of evolutionary ethics, and a look at the American pragmatist John Dewey will show how his philosophical position, potentially the most sympathetic, followed the consensus of the philosophical community in rejecting an evolutionary approach to ethics.

6. de Laguna, "Stages of the Discussion of Evolutionary Ethics," 589. See also the several critical articles in the International Journal of Ethics, such as Alfred Benn, "The Relation of Ethics to Evolution," 11, no. 1 (1900): 60–70; H. W. Wright, "Evolution and Ethical Method," 16, no. 1 (1905): 59–68; George Mead, "The Philosophical Basis of Ethics," 18, no. 3 (1908): 311–323; and Norman Wilde, "The Meaning of Evolution in Ethics," 19, no. 3 (1909): 265–283. Wilde commented in his article that evolutionary ethics assume moral evolution and provide no guide or rational justification and that "in this point lies the essential weakness of all naturalistic theories of evolution as attempted explanations of the value of conduct. They attempt to play 'Hamlet' with Hamlet left out" (283).

Popularizers

Benjamin Kidd

The most well known of the authors who used evolution as a starting point in their consideration of society was Benjamin Kidd. His *Social Evolution* (1894) received an enthusiastic reception and caused a sensation on both sides of the Atlantic.[7] Kidd, who was inspired by both Darwin and the German neo-Darwinian August Weismann, set out on an ambitious project to reconcile religion and science as well as to promote reform without encouraging socialism. Competition, he stressed, was the central progressive force in nature and in culture. Unlike those who contended that competition was an important factor in the history of man but would in time give way to cooperation, Kidd argued that only the conditions of competition would (or should) change. Any lessening of competition would inevitably lead to degeneration. Of Spencer's vision—an ideal society in which enlightened self-interest guided conduct and a conciliation of interests existed—he wrote,

> The evolutionist who has once realized the significance of the supreme fact up to which biology has slowly advanced,—namely, that every quality of life can be kept in a state of efficiency and prevented from retrograding only by the continued and never-relaxed stress of selection—simply finds it impossible to conceive a society permanently existing in this state. He can only think of it existing at all on one condition—in the first stage of a period of progressive degeneration.[8]

Kidd was not advocating a view of society based on a simple analogy with natural selection. Although he believed that selection operated on human groups to favor those with progressive organization, he argued that the main integrative force in history was religion. Moreover, he called attention to an ambiguity in the Darwinian position, the inevitable conflict between the individual and

7. For an excellent discussion on Kidd see: D. P. Crook, *Benjamin Kidd: Portrait of a Social Darwinist* (Cambridge: Cambridge University Press, 1984). Also valuable are Jones, *Social Darwinism and English Thought*; Richard Hofstadter, *Social Darwinism in American Thought* (Philadelphia: University of Pennsylvania Press, 1944); and Robert Bannister, *Social Darwinism: Science and Myth in Anglo-American Social Thought* (Philadelphia: Temple University Press, 1979).
8. Benjamin Kidd, *Social Evolution* (New York: Macmillan, 1894): 292.

society. Kidd argued that it was a tension that could not be resolved, channeled, or alleviated rationally. The salvation of man,
instead, was through the evolution of a religious impulse that sub-
ordinated the interests of the individual to that of the group. It
was religion that formed the foundation for ethics and was responsible for the progress of man, past and future.

In his later writings, Kidd elaborated a new Christianity that
opposed both unfettered capitalism and socialism. Although he
held that "philanthropic imperialism" would pave the way for a
future peaceful world order, he argued that uncontrolled competition would not of itself lead to social harmony; rather, it would
have to be restrained by ethical principles.[9] He was, in fact, critical
of the "bloom of competition," as the famous American Social
Darwinist William Graham Sumner referred to the millionaires
of nineteenth-century capitalism. Like Huxley, Kidd opposed such
"Social Darwinism." He also was alarmed by the eugenic program
espoused by Darwinists such as Francis Galton and Karl Pearson.
Their emphasis was on the importance of an individual's hereditary material and the potential for wise selection. Kidd believed
that approach to be fundamentally misguided for two reasons: it
mistook the level of human evolution to be at the individual level,
and its elitist thrust gave no hope to the mass of mankind. Instead,
Kidd stressed the sociocultural evolution of mankind, which he believed was an evolution of society, not of individuals, and which
rested on a spiritual development rather than on a rational mas-
tery of nature. In his last work, The Science of Power (1918), Kidd
claimed that the future progress of man depended on psychic evolution, which operated at a rate dramatically faster than natural
selection. Unlike the earlier, slower, and less efficient natural selection of individuals, which could be seen in the development of
Western civilization's aggressive masculine cultures, the future belonged to the power of women, who, unlike the short-range, individualistic, and intellectual "fighting male," had at the core of their
being an instinct for "duty, sacrifice, and renunciation."[10] "Woman
is indeed the actual prototype of all the great systems of religion,

9. Crook, Benjamin Kidd, 156–157.
10. Benjamin Kidd, The Science of Power (New York and London: Putnam's
Sons, 1918): 205.

of morality, of law, upon which integrating civilization rests."[11] By her evolution, woman had developed the psychological necessity of subjugating the present for the future and valuing social integration of individual advancement. Kidd stressed the difference between social and physical evolution and argued that human progress depended on the former. Human progress, therefore, could be very rapid. In the shadow of the First World War, Kidd optimistically claimed (not uniquely in this century), "Give us the Young and we will create a new mind and a new earth in a single generation."[12]

The social development of man, according to Kidd, was based on the emotional validity of moral values, chiefly duty, sacrifice, and concern for the future well-being of the race over the current advantage to the individual. This position took Kidd very far from the Darwinian one in evolutionary ethics. Along with Huxley, he saw a conflict between human values and the natural selection process and looked to the human heart for guidance. But what he saw differed from what Huxley felt or what Spencer hoped would be felt. Kidd's writings appealed to a different set in the period bracketing the Great War: those who feared the rising tide of the labor movement but did not identify with or condone the robber barons; those who admired the German emperor, who in his annual addresses to the military on the occasions of the annual swearing in demanded "sacrifice, duty, discipline, devotion, [and] iron obedience in the service of the national ideals,"[13] but were horrified by the carnage of the Great War; and those who recognized the intellectual worth of science but longed for a legitimate standing for emotions. Kidd appealed to them and provided them with an acceptable, if not altogether consistent, synthesis. Although he stressed Darwin's intellectual contributions—indeed the Origin of Species was presented in The Science of Power as one of the major intellectual shifts in Western culture—Kidd ultimately argued that Darwinian evolution with its stress on natural selection failed to provide an adequate basis for ethics. For Kidd, human evolution and progress followed different rules, and while biological evolution could tell us a lot about the past behavior of man, it was not a guide for the future.

11. Ibid., 207.
12. Ibid., 309.
13. Ibid., 141.

Woods Hutchinson

Seemingly more closely tied to the spirit of Darwinism was Woods Hutchinson, a physician and popular American medical writer and author of The Gospel According to Darwin (1898). Hutchinson wrote a series of books at the turn of the century advocating a modern, secular, "natural" approach to health, which stressed the value of fresh air, cold sponge baths, exercise, sanitation, sunshine, and common sense. He abhorred follies (old or modern) such as corsets, fashionably tight shoes for women, and the new aesthetic of slenderness ("fat is nature's savings bank").[14] He likewise opposed the various health food fads of the time as well as body-building exercises.[15] As a champion of modernism, he argued that the alleged "new diseases" of civilization were the result of old diseases and barbaric conditions (chiefly, lack of proper sanitation). To the claim that slums and overcrowding bred disease, his response was, "The remedy for the evils of civilization is more civilization!"[16] Engineering, government inspection, education, and industrialization were leading man to a higher level of progress than ever before.

The traditional religious views of man were, for Hutchinson, one of the obstacles to further progress. The early history of Western religion produced a duality in the conceptions of good and evil with each deified or personified into a divine, immortal being. More recently, man prayed to the power of good to protect him from the power of evil. However, the devil, "in spite of his fallen and degenerate condition," paradoxically had exerted a fascination for the pious. "So incessant and tremendous is the struggle to escape his clutches, that one can hardly help wondering whether he has not practically become the real object of worship to the shivering and self-tortured monk, the Jesuit with his torch and rack, the beauty-hating, witch-burning Puritan, or the modern camp-meeting exhorter with his hell-fire and brimstone. Judged by their

14. Woods Hutchinson, Common Diseases (Boston: Houghton Mifflin, 1913): 48.

15. For a very good discussion of Hutchinson and his context, see James C. Whorton, Crusaders for Fitness: The History of American Health Reformers (Princeton: Princeton University Press, 1982).

16. Woods Hutchinson, Civilization and Health (Boston: Houghton Mifflin, 1914): 13.

frenzied excesses and their fruits, Satan, rather than Jehovah, is their God."[17]

Hutchinson looked to evolution for moral guidance. "The Gospel according to Darwin" rings out the trumpet call, "There is no God but The Good."[18] He argued that our moral sense was a natural instinct that had developed in time. Although he regarded nature unsentimentally and described the importance of pain (and its avoidance), the value of courage ("the mother of the virtues"), and the selective benefit of the institution of prostitution ("a crematory, into which are hurled the least desirable elements of both sexes"), Hutchinson did not rely on natural selection for the foundation of morality but regarded human society as developmentally progressive. Moreover, like Stephen, Spencer, and Fiske, he believed that cooperation replaced competition in higher societies. Moral codes were relative to time and place and had their origin not in a utilitarian calculus of self-interest but in the social instincts and sympathies that arose out of the biological development of early man and his family.

Hutchinson's Darwinism was confined to explaining the origins of moral and social instincts. He also expanded in a Darwinian fashion on other natural instincts of man which society occasionally perverted or repressed (much to man's detriment), such as our reproductive instincts or children's affinity for sweets. Once past the origin of our instincts, however, Hutchinson primarily floated freely in his own value system. His evolutionism, then, although foundational, was not systematic or very profound. It permitted him a secular starting point from which to argue for the importance of those values he held dear. His espousal of a progressive development of man gave him the excuse to argue for change and against what he took to be retrogressive policies that impeded progress (such as the women's suffrage movement, which he believed downgraded women's sexual role in favor of higher education).

---

17. Woods Hutchinson, The Gospel According to Darwin (Chicago: Open Court, 1898): 24. For a more sympathetic view at the time of the importance of the devil, see the anonymously published work by George Millin, Evil and Evolution: An Attempt to Turn the Light of Modern Science on to the Ancient Mystery of Evil (London: Macmillan, 1896).

18. Hutchinson, Gospel According to Darwin, 25.

Samuel Alexander

One hundred eighty degrees removed from Hutchinson was the Jewish, Australian, Oxford-educated philosopher at Manchester, Samuel Alexander. Alexander had studied at Balliol College where German idealism had been introduced by Benjamin Jowett and popularized by T. H. Green and others. Jowett may have been responsible for giving Hegel a hearing at Oxford, but he was highly selective in what he accepted.[19] Alexander recounted later that Jowett once responded to his statement that he had just completed reading all of Hegel over a long vacation by commenting, "It's a great thing to have read the whole of Hegel; but now that you have read him, I advise you to forget him again."[20] And although Alexander claimed that he took Jowett's advice, even a cursory reading of Alexander's writings reveals a Hegelian spirit underlying the text. Or, at least, an Anglo-Hegelian perspective. Just as Green took from Hegel what he admired and left what he did not in elaborating an ethical position, so, too, did Alexander borrow from the idealist philosophy of his mentors at Oxford. That his later philosophy bore little resemblance to German idealism does not negate his early inspiration, or the family resemblance to what has been called the "Anglo-Aristotelian-Hegelian movement in British ethics."[21]

Alexander did not believe in the existence of a collective mind that unfolded in time, or that reality was a cosmic dream occurring in God's mind. He did, however, think of the universe in pantheistic terms and held that categories of thought were objective relations.[22] He greatly admired Spinoza and came close to the naturalism of that philosopher who also was deeply concerned with ethics. But unlike the static conception of Spinoza, Alexander believed that the cosmic process was dynamic and that values emerged

---

19. For an interesting discussion of Jowett's Hegelianism, see Peter Hinchliff, Benjamin Jowett and the Christian Religion (Oxford: Oxford University Press, 1987).

20. Alexander, Philosophical and Literary Pieces, 6. The quotation is from the introductory memoir by John Laird. On Alexander's intellectual development, also see Michael A. Weinstein, Unity and Variety in the Philosophy of Samuel Alexander (West Lafayette: Purdue University Press, 1984).

21. Alexander, Philosophical and Literary Pieces, 19.

22. Bertram Brettschneider, The Philosophy of Samuel Alexander: Idealism in "Space, Time, and Deity" (New York: Humanities Press, 1964): 34.

from it. "Herein lies for philosophy the significance of Darwinism," he stated in an essay entitled "Naturalism and Value." Although values were relative to men, they were not arbitrary but founded in the nature of things.[23] They did not have a separate Platonic existence but were embodied in natural objects or their relations. Darwinism was the natural history of value, according to Alexander, in that

> in human affairs the values prevail and establish themselves by ostracizing the unvalues. Virtue . . . imprisons vice or otherwise makes its life difficult. The moderately good enforce their own tastes on the moderately bad, and compel the bad to at least apparent conformity. Truth in the end, after long years, drives out error (though not ignorance). Beauty drives out ugliness; in the end it is the beautiful imaginations which remain. The separation of value and unvalue is always a matter of experiment or trial, and it is a struggle of opinions or sentiments which uses the forces of persuasion or tradition.[24]

Darwinism meant for Alexander that "successful and permanent types of life in the plant and animal world are those which can prevail under the conditions of existence over real or possible rivals."[25] The debates over particulars of the theory seemed unimportant to Alexander. "So far as natural selection operates it secures the success . . . of certain types to the exclusion of other types."[26] He interpreted natural selection to mean that in nature there was a selective process that assured preference to superior permanence. In the human world this took the form of establishing values that by their superiority replaced "unvalues."[27]

Alexander, of course, violated the spirit of Darwinism. Relative fitness for Darwinists did not mean that given certain environments particular relations or variations would necessarily prove superior. There was, according to Darwinism, an element of chance in nature, and fitness was never perfect. Concepts like "worthy" or "superior," therefore, had no legitimate place in biological thought. Although Darwin spoke of the progress of man, he did not hold that it was inevitable but rather an empirical fact that could be explained by evolution. Alexander's emergent philosophy

23. Alexander, *Philosophical and Literary Pieces*, 283.
24. Ibid., 286.
25. Ibid.
26. Ibid.
27. Ibid.

suggested otherwise. The values that emerged in time must domi-
nate in time. Values were not defined by success, but succeeded.
They did so because there was an overall direction to the process,
and in Space, Time, and Deity (the Gifford Lectures at Glasgow)
Alexander sketched a cosmic system striving toward perfection.
Darwinism was reduced to meaning that value, when it emerged,
survived ultimately because of its greater perfection.

Alexander was no closer to Spencer. Although Alexander's teleology was more in the style of Spencer, his pantheism was not.
Nor did he stress the active adaptation of life to changing conditions, which was such a vital part of Spencer's philosophy.

Kidd, Hutchinson, and Alexander reflected the evolutionary perspective on morality at the turn of the century. Like the fascinat-
ing but vulgar mansions of the robber barons of the period, these
three authors constructed intellectual structures that from the vantage of the twentieth century are curious amalgams. Although popular at the time, they failed to garner any professional support, and
when tastes shifted, they were left largely unremembered by the
next generation.

Critics

British Philosophy and G. E. Moore

Alexander was unusual in his espousal of an evolutionary system, not so much because of its originality, for we have seen that
numerous teleological systems of evolutionary ethics were devised
in the late nineteenth century, but rather because professional philosophers had pretty much confined such systems to the realm of
"popular" literature. Alexander's peers at Oxford were hostile to
Spencerian or Darwinian ethics, and philosophers at Cambridge,
England, Cambridge, Massachusetts, Oxford, and Chicago were
in agreement at least on that score.

Time did little to change the philosophers' stand. William Ritchie
Sorley, who replaced Sidgwick as Knightbridge Professor of Moral
Philosophy at Cambridge, held a critical opinion of evolutionary
ethics. Before teaching at Cambridge, he published The Ethics of
Naturalism: A Criticism (1885), which was an extended examination of "the ethical significance of the theory of evolution." His
conclusion was "that the theory of evolution—however great its

achievements in the realm of natural science—is almost resultless in ethics."[28] Like Sidgwick, Sorley held that evolutionary theory could not supply a perspective from which to resolve critical ethical issues like those of the conflict between the individual and society. Sorley not only agreed with Sidgwick that ethical naturalism—that is, those systems that started from an empirical basis—failed but also followed him in the belief that a religious assumption was necessary for any valid ethical system. Sorley, however, went beyond his predecessor in elaborating a theistic philosophy as a new foundation for ethics. Like Sidgwick's attempt to save utilitarianism by postulating an initial divine sanction, it was not widely regarded as successful.

Of greater professional importance at Cambridge in the generation after Sidgwick was G. E. Moore, who repeated the condemnation of evolutionary ethics. Moore's arguments were important because they were embedded in a broader position that defined a basic starting point for twentieth-century ethics.

In his classic work, Principia Ethica (1903), Moore was careful to distinguish Darwin's ideas from those of Spencer and his followers. He referred to Darwin as a major figure in biological thought and classified Spencer as the best known of the many popular writers on "Evolutionistic Ethics." Moore devoted ten pages to criticizing Spencer's writings on ethics, which he treated like a set of inferior undergraduate essays. In commenting on Spencer's Data of Ethics, Moore's contempt was palpable.

> It is, of course, quite possible that his treatment of Ethics contains many interesting and instructive remarks. It would seem, indeed, that Mr Spencer's main view, that of which he is most clearly and most often conscious, is that pleasure is the sole good, and that to consider the direction of evolution is by far the best criterion of the way in which we shall get most of it: and this theory, if he could establish that amount of pleasure is always in direct proportion to evolution and also that it was plain what conduct was more evolved, would be a very valuable contribution to the science of Sociology; it would even, if pleasure were the sole good, be a valuable contribution to Ethics. But the above discussion should have made it plain that, if what we want

28. William Ritchie Sorley, The Ethics of Naturalism: A Criticism (Edinburgh: William Blackwood and Sons, 1904): 309. Also see his Recent Tendencies in Ethics: Three Lectures to Clergy Given at Cambridge (Edinburgh: William Blackwood and Sons, 1904) and Moral Values and the Idea of God (Cambridge: Cambridge University Press, 1918).

from an ethical philosopher is a scientific and systematic Ethics, not merely an Ethics professedly "based on science"; if what we want is a clear discussion of the fundamental principles of Ethics, and a statement of the ultimate reasons why one way of acting should be considered better than another—then Mr Spencer's "Data of Ethics" is immeasurably far from satisfying these demands.[29]

Although Spencer was a ripe target for Moore's analytical attack, his criticism of evolutionary ethics extended to all variants. Moore distinguished a number of possible approaches to evolutionary ethics in his Principia Ethica. For example, he noted that if we accepted the ethical judgment (independent of Spencer) that evolution was "progress," that is, the more evolved was indeed the better, we had little to go on in determining the ethical value of any particular act. "We cannot assume that, because evolution is progress on the whole, therefore every point in which the more evolved differs from the less is a point in which it is better than the less."[30] And, "we certainly cannot use it as a datum from which to infer details."[31] That is, even if a cosmic process leading to progress on the whole existed, it did not supply a point from which we could derive any knowledge of particular moral judgments.

The more common view in evolutionary ethics, according to Moore, was a simpler claim: "that we ought to move in the direction of evolution simply because it is the direction of evolution."[32] Moore claimed this view was invalidated because it was an example of the "naturalistic fallacy—the fallacy which consists in identifying the simple notion which we mean by 'good' with some other notion."[33]

Moore's Principia Ethica is best remembered for its general assault on all ethical systems based on the naturalistic fallacy. These "naturalistic ethics," which he claimed included the major versions of evolutionary ethics as well as over half the writers of Western ethics,[34] attempted to define the "good" by reference to some

29. George Edward Moore, Principia Ethica (Cambridge: Cambridge University Press, 1978): 54.
30. Ibid., 55.
31. Ibid., 55.
32. Ibid., 56.
33. Ibid., 58.
34. See Thomas Baldwin, G. E. Moore (London: Routledge, 1990). Baldwin has a good discussion of Moore's critique of earlier ethical positions.

"other thing."[35] This other thing might be a natural object, or an object inferred to exist in some "supersensible real world." Most of British philosophical writing on ethics fell into one or the other of these two categories. Evolutionary ethics, especially Spencer's, was, for Moore, among the least formidable of his opponents. And in developing his critique of naturalistic ethics he was primarily attempting to show the inadequacy of Mill, Bentham, McTaggart, Bradley, and Green. In their place Moore set out to demonstrate that "good" was a simple notion that could not be defined or analyzed. It therefore could not be defined as pleasure, which the utilitarians contended, or as an evolutionary adaptation. Its existence was simply apprehended.

Moore's writings had the dramatic effect of redirecting much ethical discussion, and over the years an enormous body of literature has been generated in response to his position.[36] A recent study of Moore claims that "the influence of Moore's ideas spread well beyond Cambridge (and Bloomsbury), partly through the establishment of an 'intuitionist' school of ethical theory at Oxford which largely agreed with him on the central metaphysical issues of ethics. The result is that, for better or worse, twentieth-century British ethical theory is unintelligible without reference" to the Principia Ethica.[37] The author of this statement meant by his claim that Moore's position "was thought to make unacceptable metaphysical and epistemological demands; so the only recourse was to abandon belief in an objective moral reality and accept an emotivist, prescriptivist, or otherwise anti-realist, account of ethical values."[38]

Later philosophers, although they rejected Moore's positive contributions to moral philosophy, followed him in criticizing the metaphysical systems of earlier philosophers like Green and in continuing to dismiss Spencer and other writers on evolutionary ethics. Moore's Principia Ethica is, indeed, often cited as having sunk

35. Moore, Principia Ethica, 38.
36. For an interesting discussion of Moore's immediate historical impact, see Tom Regan, Bloomsbury's Prophet: G. E. Moore and the Development of His Moral Philosophy (Philadelphia: Temple University Press, 1986): 198. Also see Baldwin, G. E. Moore; Harrod, The Life of John Maynard Keynes; and Paul Levy, G. E. Moore and the Cambridge Apostles (New York: Holt, Rinehart and Winston, 1979).
37. Baldwin, G. E. Moore, 66.
38. Ibid.

evolutionary ethics "into oblivion."[39] This is a statement that, if true, certainly needs to be qualified, for we have seen that most earlier philosophers were skeptical about the ethical conclusions of Stephen, Clifford, Spencer, Fiske, and similar writers. Moore was not responsible so much for a set of new objections that led to the demise of evolutionary ethics as he was for reinforcing the attitude with which professional philosophers approached it. If Moore was better remembered than his teacher, Sidgwick, from whom he heard most of the telling arguments against evolutionary ethics, it was a reflection of Moore's greater importance in British philosophy during our century.[40] Moore captivated the Bloomsbury set who knew him as a young man, and he exerted a gentle but guiding hand in redirecting British ethics. Perhaps another reason his influence was greater than Sidgwick's was a consequence of Moore having a general program for philosophy that he articulated with consummate skill. Sidgwick, in contrast, provided a brilliant analysis of the methods of ethics but was never satisfied that he had successfully achieved an acceptable rational synthesis of our intuitive moral convictions and a rational structure to relate them.

American Philosophy and John Dewey

On the American side of the Atlantic, toward the end of the nineteenth century philosophy began to follow its own lines of development rather than merely construct variations of British empiricism or literary versions of German idealism. The intellectual world in the United States had matured to the point where a new professionalism had an effect that was similar to what was happening in Britain, that is, new professional standards led to systematic philosophy that could find little of value in "popular writings" on the importance of evolution for ethics. The great figures in American philosophy—William James, Josiah Royce, John Dewey, George Santayana—were unanimous in their rejection of Spencer and in their rejection of evolution as a guide for ethics.

William James, for example, like all the American pragmatists,

39. For example, Anthony Quinton, "Ethics and the Theory of Evolution," in I. T. Ramsey, ed., Biology and Personality: Frontier Problems in Science, Philosophy and Religion (Oxford: Basil Blackwell, 1965): 107.
40. The undergraduate lectures that Moore heard were posthumously printed as Sidgwick, Lectures on the Ethics of T. H. Green, Mr. Herbert Spencer, and J. Martineau (see fn. 33, chap. 5).

back to Chauncey Wright, interpreted ideas by their practical con-sequences. He stressed the importance of evolution for under-standing the living world and mind. But as much as he was influ-enced by an evolutionary perspective, James rejected any notion that it could provide a guide for ethical decision or a foundation for morality.[41] After a brief flirtation with Spencer's teleological ethics early in his career, he criticized it in a review for the Nation: "We can never on evolutionist principles altogether bar out per-sonal bias, or the subjective method, from the construction of the ethical standard of right."[42] He went on to explain, "For if what is right means what succeeds, however fatally doomed to succeed that thing may be, it yet succeeds through the determinate acts of determinate individuals; and until it has been revealed what shall succeed, we are all free to 'go in' for our preferences and try to make them right by making them victorious."[43] James was no more sympathetic to the Darwinian approach, and in the same year as his early criticism of Spencer (1879) he reviewed Clifford's Lec-tures and Essays for the Nation and took a swipe at the latter's se-lectionist perspective: "The entire modern deification of survival per se, survival returning into itself, survival naked and abstract, with the denial of any substantive excellence in what survives, except for more survival still, is surely the strangest intellectual stopping-place ever proposed by one man to another."[44]

William James never developed a systematic ethics; however, he wrote extensively on the subject, and it was central to his more general philosophy. In his writings his opposition to evolutionary

41. On James's relationship to evolution, see Philip Weiner, Evolution and the Founders of Pragmatism (Cambridge: Harvard University Press, 1949); Richards, Darwin and the Emergence of Evolutionary Theories of Mind and Behavior; and Gerald E. Myers, William James: His Life and Thought (New Haven: Yale Uni-versity Press, 1986). Weiner also has an interesting discussion of Chauncey Wright, as does Edward H. Madden, Chauncey Wright and the Foundations of Pragmatism (Seattle: University of Washington Press, 1963). Wright was critical of Spencer but held Darwin in high regard. Wright died at the young age of 45 and did not leave any essays on ethics. His opinions have been reconstructed from letters and a few published remarks. From these it appears that he was skeptical of attempts like Spencer's to create an evolutionary ethics but that he gave his own moral thoughts, which were based on utilitarianism, an evolutionary twist on occasion. See J. J. Chambliss, "Natural Selection and Utilitarian Ethics in Chauncey Wright," Amer-ican Quarterly 12 (1960): 144–159.

42. James's remark, which originally was published in 1879, appears in his Collected Essays and Reviews (London: Longmans, Green, 1920): 148.

43. Ibid., 148–149.

44. Ibid., 143–144.

ethics was clearly expressed.[45] And even a cursory reading of the other pragmatists reveals little to suggest that any were more sympathetic. John Dewey, the one figure of the pragmatist school who because of his emphasis on the importance of Darwin for philosophy might have been expected to be a champion for evolutionary ethics, was, like James, openly critical.

Dewey appreciated the impact of Darwin's ideas and often stressed the importance of evolution for an understanding of man. He valued historical research and was stridently naturalistic. But Dewey was biting in his critique of both Spencerian and Darwinian attempts to derive ethical maxims from evolution. In his highly influential textbook, Ethics (1908), which he wrote with his fellow pragmatist colleague James Tufts, such attempts were dismissed as pseudo-science and as a parody of the facts.[46] The view that a biological perspective could clarify the domain of human ethics was treated as a fundamentally misguided notion: "The chief objection to this 'naturalistic' ethics is that it overlooks the fact that, even from the Darwinian point of view, the human animal is a human animal."[47] Erecting central principles, such as efficiency or achievement, from a reading of nature, according to Dewey and Tufts, was taking means for ends, a fallacy common to all materialism.[48]

This is not to say that Dewey regarded the theory of evolution as uninteresting for the consideration of ethics. A decade earlier he had discussed the topic of evolutionary ethics in reference to Huxley's Romanes Lecture. In Dewey's perceptive article, he noted how philosophical subjects rarely were pursued to conclusive ends but rather in midstream the interests of scholars shifted and went off in other directions. Dewey applied this perspective to the history of the early debates on evolutionary ethics and contended that

45. On James's philosophy, see the carefully crafted biography by Myers, William James, which replaces the earlier two-volume biography by Perry as the major study on James. See Ralph Barton Perry, The Thought and Character of William James (Boston: Little, Brown, 1935). Also of interest are Graham Bird, William James (London: Routledge and Kegan Paul, 1986); Bernard Brennan, The Ethics of William James (New York: Bookman Associates, 1961); John K. Roth, Freedom and the Moral Life: The Ethics of William James (Philadelphia: Westminster Press, 1969); and Ellen Kappy Suckiel, The Pragmatic Philosophy of William James (Notre Dame: University of Notre Dame Press, 1982).

46. John Dewey and James H. Tufts, Ethics (New York: Henry Holt, 1908): 369.

47. Ibid., 372. The quotations from Ethics appear in the sections written by Dewey.

48. Ibid., 373.

they commenced with the issue of the relationship of man to the lower animals, that is, whether or not a chasm existed between man's intellectual, moral, and physical nature and that of lower forms. A shift in interest, dating from Huxley's Romanes Lecture, allegedly redirected the topic to a discussion of the relationship of evolutionary concepts and ethical concepts.[49] The accuracy of Dewey's claim was debatable, for several important studies tracing the moral sentiment from animal to man dated from the 1890s, for example, Alexander Sutherland's Origin and Growth of the Moral Instinct[50] and Henry Calderwood's Evolution and Man's Place in Nature. David Ritchie, author of the popular Darwinism and Politics, claimed that the origin of the moral sense "forms the best initial test of the adequacy or inadequacy of the theory of natural selection outside the merely biological domain."[51] And as late as 1915, L. T. Hobhouse published his Mind in Evolution, which reflected the continuing discussion on the origin of the moral sentiment that went on in journals, particularly those that stressed psychological issues, such as Mind.[52]

Although Dewey's point may have lacked historical veracity, it nonetheless did point to an important distinction: that the location of the historical origins of the moral sentiment was not necessarily the discovery of a foundation for an ethical system. The former question was what Darwin concerned himself with as an issue that had relevance for the acceptance of a general evolutionary perspective on the living world. The latter concerned those writers, some Darwinian, some not, who were searching for a new system of ethics and saw in the natural world a potential foundation. Dewey may have gotten his historical facts a little jumbled in portraying these issues in a chronological sequence, but that they were distinct issues is worth noting and has been stressed repeatedly by critics of evolutionary ethics.

Although Dewey agreed with Huxley that the theory of evolution did not provide an adequate guide for action, he chided Hux-

---

49. John Dewey, "Evolution and Ethics," The Monist 8, no. 3 (1898): 322.

50. Alexander Sutherland, The Origin and Growth of the Moral Instinct, 2 vols. (London: Longmans, Green, 1898).

51. David G. Ritchie, Darwinism and Politics: With Two Additional Essays on Human Evolution, 3d ed. (London: Swan Sonnenschein, 1895): 96. The quotation is from an essay published originally as "Natural Selection and the Spiritual World," Westminster Review 133 (1890): 459–469.

52. L. T. Hobhouse, Mind in Evolution (London: Macmillan, 1915).

ley for overstating his case and argued that evolution was an important background for philosophy. Dewey accepted Huxley's rejection of crude "Social Darwinist" formulations that exalted ruthless selfishness. But, unlike Huxley, he did not frame the relationship of man to nature as a cosmic conflict. Rather, Dewey stressed continuity but contrasted the differences between biological evolution and social progress. If there was an analogy to be made between the process of speciation and the advancement of society, it was in the creation of new paths of development. There may be, according to Dewey, a selection operating in human culture, but it was not the crude Malthusian one that Darwin depended on in his description of the origin of species. Instead Dewey pointed out the conscious and deliberate form of action that man could and did take in response to his goals. Public opinion and education encouraged and discouraged types of action.[53] The animal instincts that we have inherited were, therefore, no more than "promptings" and were "no more sins than . . . saintly attributes."[54] It was in the conscious training and directing of our "animal" nature that ethical issues intruded. Evolution, then, served for Dewey as a backdrop. It explained how it was that man had appeared on earth and served as a naturalistic bulwark against importation of religious and metaphysical ethical demands. But it was not a foundation for ethics. Natural selection could not be generalized into a criterion to justify actions that promoted the general good, nor could it serve to sanctify actions that were leading toward an evolutionary goal.

Dewey's theory of value, which was later in the century overwhelmed by analytical philosophy and is only now making a minor comeback in professional philosophy, approached ethics in a manner that was very different from that of the supporters of evolutionary ethics. Although he recognized the "cosmic roots" of morality in custom, ritual, and so on, Dewey argued that a complete morality existed only when an individual freely chose the good. Moreover, such choice was not judged by a set of rules or criteria such as actions leading to the greatest good or survival value for society.

Dewey's values reflected a mainstream American liberal reform stance. He, however, did not take for granted the received moral

53. See Dewey's "Evolution and Ethics."
54. Ibid., 330.

precepts of his milieu. Dewey stressed the dynamic nature of society and did not accept the smug Victorian notion that "right" and "wrong" were obvious. He stood in marked contrast to writers like Jacob Gould Schurman, who in his Ethical Import of Darwinism (1887) wrote that a historical study of ethical positions would clarify what men have everywhere and at all times considered right and wrong. As Schurman noted, "All are agreed that certain courses of conduct are right and the opposite wrong, moralists seem unable to agree in anything except the contradictory claim of building their incompatible theories upon these universally recognized propositions."[55]

Dewey did not attempt to build his theory on such allegedly recognized propositions. Instead he stressed the unique character of all moral decisions. Moral inquiry, like all other inquiry, began with the recognition of a situation that was problematic. Such situations were always complex contextual wholes and required careful evaluation. Inquiry, whether about the natural world or about a moral dilemma in the social world, took an indeterminate, problematic situation and formulated a testable hypothesis that proposed a solution judged adequate to the situation at hand.[56]

The Darwinian world served Dewey as a background. He believed that the organic world constantly adjusted to its changing environment and in so doing modified the system of which it was a part. Although Dewey had little use for the simpleminded application of natural selection as a criterion of human action, there was a weak sense in which Dewey's conception of moral inquiry had a Darwinian cast: he thought of "truth" as those conditions that resolved a problematic situation. But fitting Dewey simplistically into an evolutionary mold does violence to his philosophy. To be sure, the ideas that we judged adequate were ones that would stand up to severe scrutiny, but far from a blind selectionist process, human actors consciously made decisions within a social

55. Jacob Gould Schurman, The Ethical Import of Darwinism, 3d ed. (New York: Charles Scribner's Sons, 1887): 2. Schurman was critical of evolutionary ethics. His ideas are more in the tradition of comparative culture that led to approaches like those of ethical relativity. See, for example, Edward Westermarck, The Origin and Development of Moral Ideas, 2 vols. (London: Macmillan, 1906–1908), and his more accessible Ethical Relativity (London: Kegan Paul, Trench, Trubner, 1932).

56. For a sympathetic and insightful discussion, see H. S. Thayer, Meaning and Action: A Critical History of Pragmatism (Indianapolis: Bobbs-Merrill, 1968).

framework that was far removed from the calculus of pleasures or the calculus of population genetics.

The decades flanking the turn of century, then, were ambivalent times for evolutionary ethics. Popular writers, indulging in speculative social philosophy, spun out various versions of ethics with alleged ties to the theory of evolution, or to more general developmental philosophies. Such views, like those of Kidd, often had brilliant, if brief, moments of public interest, but none were able to penetrate into the halls of academia or into the serious intellectual journals. No proponent was able to clothe evolutionary ethics in acceptable or rigorous language to bring it successfully into the twentieth century. But the story was far from over.

7. Syntheses, Modern and Otherwise

1918–1968

By the conclusion of the Great War, evolutionary ethics was approaching a dead end. Philosophers had abandoned discussion of evolutionary approaches to ethics and disdainfully ignored "ill-conceived" attempts to revive them. Professionals in other disciplines that treated man's moral sentiment and its expression had shifted their perspectives away from evolutionary explanations and were now asking different questions and using different methods to answer them. Anthropology, for instance, focused less on developmental schemes and concepts than it had in the last third of the nineteenth century. Even before the war, the German Franz Boas, who settled in the United States and was a major influence in the definition of the relatively new field of anthropology here, espoused in his Mind of Primitive Man (1911) a more relativistic conception of culture than earlier American social evolutionists such as Lewis Henry Morgan or John Wesley Powell.[1] Under Boas's influence, American anthropology came to stress the psychological dimension of culture and the individual's relationship to it.[2] In Britain, a different but similarly less evolutionary approach dominated anthropology through the research and writing of "func-

1. See George W. Stocking, Jr., ed., The Shaping of American Anthropology 1883–1911: A Franz Boas Reader (New York: Basic Books, 1974); Stocking, Race, Culture and Evolution: Essays in the History of Anthropology (New York: Free Press, 1968); Degler, In Search of Human Nature; John S. Haller, Outcasts of Evolution: Scientific Attitudes of Racial Inferiority, 1859–1900 (Urbana: University of Illinois Press, 1971); William Culp Darrah, Powell of the Colorado (Princeton: Princeton University Press, 1951); Curtis M. Hinsley, Jr., Savages and Scientists: The Smithsonian Institution and the Development of American Anthropology, 1846–1910 (Washington, D.C.: Smithsonian Institution Press, 1981); Bernhard J. Stern, Lewis Henry Morgan: Social Evolutionist (Chicago: University of Chicago Press, 1931); Carl Resek, Lewis Henry Morgan: American Scholar (Chicago: University of Chicago Press, 1960); and John Zernel, "John Wesley Powell: Science and Reform in a Positive Context," Ph.D. dissertation, Oregon State University, 1983.

2. See Stocking, Victorian Anthropology, 287.

tionalists" such as Bronislaw Malinowski who sought to bring the insights of psychology to anthropology to understand the function of social practices in a cultural context. Other functionalists, like A. R. Radcliffe-Brown, stressed the comparative study of social structures.

The shift away from an evolutionary perspective was evident in psychology also, where the emergence of behaviorism signaled a new approach to the study of mind. The experimentalism of the biological sciences led to impressive and initially fruitful results when applied to psychology, and the "evolution of mind" looked decidedly old-fashioned, a lethal trait in scientific fashion.

It is hardly surprising, then, that in spite of the extensive research on biological evolution, extensions of, or analogies with, evolutionary thought were increasingly shunned in the social sciences during the latter part of the first half of the twentieth century.[3] Instead, researchers chose approaches that they believed were more appropriate and would be more productive. The cultural disillusionment in the Anglo-American world following in the wake of the First World War served to reinforce this flight from evolutionary philosophy. The inevitability of human progress was less obvious given memories of trench warfare and mustard gas. Spencer's optimistic evolutionary philosophy, for example, no longer commanded the serious attention it had before the war. Increasingly, sets of his dated work appeared on the shelves of used book dealers where they in time became a drug on the market.[4]

Interwar Evolutionary Ethics

With the demise of evolutionary social thought, one might have expected an end to evolutionary ethics in the 1920s. But intellectual history is unpredictable. Far from disappearing, evolutionary ethics experienced a rebirth in the first half of the twentieth century among speculative thinkers who continued to seek a naturalistic foundation for a progressive view of nature and man. These

3. Carl Degler discusses the ideological dimension of this shift in his In Search of Human Nature.
4. Lack of any subsequent revival of interest has left hundreds of sets of Spencer in used book shops, to be sold primarily to historians or to the general public as inexpensive decorator items. They have become cheaper, even, than sets of Thackeray or Scott.

attempts, which differed significantly from earlier ones, and the reactions to them constitute the second episode in the history of evolutionary ethics.

Although the Victorians' faith in progress allegedly was killed off on the fields of Flanders, even a cursory reading of the popular literature of the 1920s reflects that alongside expressions of deep disillusionment and despair, there were many attempts to recast and revive a positive and inspiring faith in man's progress—actual or potential. Some biologists thought that the earlier interpretations of evolution, which stressed natural selection and competition, had been misguided both in distorting our picture of nature and in promoting deleterious courses of human action. The moral lesson was clear to these writers: German Darwinists' emphasis on natural selection and the struggle for existence had contributed to German militarism and to war. In reaction, scientists like Vernon Kellogg, David Starr Jordan, William Patten, and Edwin Conklin advocated alternative visions that stressed cooperation on both the biological and social levels.[5] Although these writings stressed moral issues, they did not propose fully developed systems of evolutionary ethics. William Patten, for instance, in his book The Grand Strategy of Evolution (1920), discussed the parallels among all living beings and associations of life.[6] He argued that it was cooperative action that characterized the living world and was responsible for its progress. Altruism was not confined to the human domain but was basic to the evolution of all living forms. The "struggle for existence" was not, according to Patten, to be interpreted in a crudely Darwinian fashion; instead it was "a struggle of the individual to find the right way out of the obstructive conditions created by its own growth and by that of other individuals, in order to give itself to a larger life."[7] This process occurred among cells of an organism, or individuals of a social group, and was the guide

5. See the very interesting and well-documented article by Gregg Mitman, "Evolution as Gospel: William Patten, the Language of Democracy, and the Great War," Isis 81, no. 308 (1990): 446–463. Mitman places this story into a wider context in his The State of Nature: Ecology, Community, and American Social Thought, 1900–1950 (Chicago: University of Chicago Press, 1992). Also see J. W. Atkinson, "E. G. Conklin on Evolution: The Popular Writings of an Embryologist," Journal of the History of Biology 18, no. 1 (1985): 31–50.

6. William Patten, The Grand Strategy of Evolution (Boston: Richard G. Badger, 1920).

7. Ibid., 55.

for proper living. The lesson was simple: mutual service "creates a larger unity, and a larger individuality out of smaller ones. Service that flows out of one individuality to another we call benevolence, or altruism. When a constructive way is found, we call it the right way, or rightness, because it is creative and preservative to that extent."[8] For Patten, man's challenge for the future was in the creation of social structures that fostered "national and international cooperation, commercial and educational, resting on some common understanding; on some basic community of methods and motives."[9]

Closer to a new evolutionary ethics were numerous "emergent evolution" theories that purported to account for the appearance of man and his unique cultural characteristics. Some were clearly religiously inspired and were a continuation of the syncretistic thinking of the previous century. C. Lloyd Morgan, the famous student of animal behavior, for example, wrote numerous works on emergent evolution. He acknowledged Samuel Alexander as a major influence,[10] but he differed from Alexander in his assessment of the relevance of natural selection for ethics. Like T. H. Huxley, Morgan separated the lower forces of speciation from the higher processes of moral development and claimed that with the emergence of man, a new level of evolution had come into being. In his early Habit and Instinct, Morgan wrote,

> No one can read the ninth volume of Huxley's "Collected Essays,"— that on "Evolution and Ethics"—without seeing that he clearly perceived the distinction between the method of natural selection and that of conscious choice which supersedes it in social evolution. The criticisms called forth by his Romanes Lecture and the reiterated assertions that he had abandoned the naturalistic interpretation of ethical phenomena, together with the defence of his position in the prolegomena prefixed to this ninth volume, all serve to indicate how essential it is that the method of conscious choice should be clearly distinguished from that of natural selection. What we strive to effect in the social evolution which embodies the results of human choice, is often very

8. Ibid., 167.
9. Ibid., 390.
10. Morgan often cited Alexander and was open in his acknowledgment. See, for example, Conway Lloyd Morgan, Habit and Instinct (London: E. Arnold, 1896): 335, and his later Gifford Lectures of 1922, which were published as Emergent Evolution (New York: Henry Holt, 1923): 9.

different from that which natural selection alone would produce. Our ideals are the products of a mental evolution which has escaped from the bondage of natural selection.[11]

Although man's moral development was not a function of natural selection, it was part of an evolutionary process, one that ultimately allowed individual human minds to glimpse the divine purpose encompassing all evolutionary advance. For Morgan, as for many inspired evolutionary writers of this period, human development was capable of reaching a stage that went beyond the acquisition of rational thought, that properly could be called a spiritual level. Morgan discussed this spiritual level in his second Gifford Lectures (1923)[12] in which he referred to the writings of the Marburg professor of theology Rudolf Otto. Otto discussed the concept of the "holy" in his widely read book, Das Heilige (1917), which was published in English translation in 1923.[13] Although Morgan did not accept the formulation of the "numinous," as Otto chose to call it, he did suggest that evolutionists, like himself, held a similar point of view. "For I, too, am now concerned to urge that what in naturalistic regard is 'epigenetic' emergence is from first to last the temporal unfolding of Divine Purpose in which there is no first nor last since it Is."[14] The divine purpose, according to Morgan, was not separate from nature, nor did he believe it was in any way personal. Rather human evolution had reached a stage where some individuals could contemplate the awesome mystery of the cosmos. For Morgan, this mystical appreciation was ancillary to religion, and it allowed him to reconcile his commitment to Christianity with his equally deep commitment to a naturalistic, evolutionary perspective.[15]

Morgan's writings were in the tradition of nineteenth-century evolutionary syntheses of religion and science, like those of Drummond and Alexander. Similar to theirs, it was a highly individual faith that merged ideas from different domains. His approach to

11. Morgan, Habit and Instinct, 335–336.
12. C. Lloyd Morgan, Life, Mind, and Spirit, Being the Second Course of the Gifford Lectures (New York: Henry Holt, 1925).
13. Rudolf Otto, The Idea of the Holy: An Inquiry into the Non-Rational Factor in the Idea of the Divine and Its Relation to the Rational (Oxford: Oxford University Press, 1923).
14. Morgan, Life, Mind, and Spirit, 308.
15. Ibid., 313.

ethics, although loosely "evolutionary," did not derive in any serious way from evolutionary insights. Evolution was part of a cosmic process that partly explained human advancement. Human values came from a contemplation of that process and an understanding of its greater purpose.

Some popular writers in the 1920s and 1930s accepted the notion of progress based on evolution but without a theological underpinning. Their ethical writings tended to depend more closely on biological evolution. Alfred Machin, for example, argued,

> If evolution is a fact, and there is a vast body of evidence which suggests that it is, and if man is a product and outcome of evolution, then an understanding of the process of evolution must explain man to himself. It must show him why he is what he is, and also why he is a member of a society which has such a distinctive structure. It must, in short, explain human nature, and show the human organism in relation to its environment in just the same way as it explains plant and animal organisms in relation to their environment.[16]

Machin did not rely on notions of emergent evolution but instead proposed a version closer to the original Darwinian formulation, which stressed the central role of natural selection. He had his own view of natural selection, however, which was based on his rejection of the Malthusian view of populations and his belief that the unit of selection was the group, not the individual.[17] Machin agreed with Darwin that natural selection operated on the animal level but insisted that human selection differed in being conscious. On both the animal and human levels, however, natural selection operated to regulate societies and their members to promote fitness. "Man is, in short, like all other living things, just a bundle of survival values," he wrote.[18]

Although Machin incorporated some of his own values into his retrospective account of the mental and moral history of man—such as his view that the main driving force of human evolution was the desire for wealth[19]—he stated that presently no adequate moral philosophy existed. So, to that extent, he was not attempting to

16. Alfred Machin, Darwin's Theory Applied to Mankind (London: Longmans, Green, 1937): xvii.
17. Machin elaborated these ideas in The Ascent of Man by Means of Natural Selection (London: Longmans, Green, 1925).
18. Machin, Darwin's Theory Applied to Mankind, 276.
19. Ibid., 193.

justify an a priori moral or political position by an appeal to evolution. He did, however, make it clear that he believed that evolution would provide the key to any future moral philosophy. "But
as yet there is no true philosophy, no rationalization which gives
effective guidance. The answers of the Greeks, Romans and of
Christianity, are outworn and unadapted to the new conditions of
this industrial age. New philosophies are in their birth throes, and
it seems most probable that evolution philosophy must furnish
that foundation which can alone give stability to any new gospel."[20] He also viewed the future with hope. In spite of predictions
that war would be "grim scientific affairs" and that the strains of
civilization would increase as a result of the replacement of physical labor by mental labor, Machin believed that progress would
continue. He stated that natural selection could not be arrested
and that we were participating in what was perhaps the greatest
drama of the universe.[21] With the memory of the Great War still
fresh in the minds of many people and the political clouds over
the continent increasingly dark, his high-minded optimism must
have appealed to those who wanted to envision a future that was
brighter than the past.

The New Psychology

The evolutionary approach to ethics found a sympathetic hearing among many in the 1920s and 1930s who wished to maintain a belief in the progress of man. Although none of them constructed a systematic evolutionary ethics, they, at a minimum, kept
the association between evolution and ethics alive. Ironically, greater
support came from a group that ultimately contributed to undermining the unquestioned belief in progress: the advocates of
the "new psychology." Recall that one of the major philosophical
questions asked by Sidgwick and other moral philosophers was,
what constituted the basis of individual moral obligation? Although evolutionary ethics might explain how concepts of moral
obligation first arose, philosophers pointed out that origins were
not the same as justifications. Some supporters of the theories
of the psychoanalytic movement claimed that psychology provided
new grounds for evolutionary ethics. In spite of its controversial

20. Ibid., 279.
21. Ibid., 283.

nature and the hostility of many academic psychologists, who increasingly were drawn to experimental questions, the psychoanalytical school suggested to a few writers that the dated evolutionary ethics of Leslie Stephen or Herbert Spencer might be reworked to yield a satisfactory evolutionary perspective on human morality. Not that the implications of psychoanalysis pointed to an agreed upon moral interpretation. A. G. Tansley, the Cambridge botanist whose New Psychology and Its Relation to Life (1920) was one of the leading popularizations of the time, made a clear distinction between psychological influences and the "higher moral self." Drawing on Freud's concepts, Tansley attempted to show how the individual was driven by unconscious causes that could lead him to rationalize selfish actions. Tansley also drew on the widely read work of Wilfred Trotter, Instincts of the Herd in Peace and War, 1916–1919 (1916). Trotter, a physician who had an interest in social psychology, held that although Freud was a pioneer in the scientific analysis of human psychology, he had not given enough attention to an instinct of central importance: gregariousness. It was this instinct, which was essential in understanding social animals, Trotter claimed, that had given rise to the sense of duty among individuals in human society. The conflict between our sense of duty and the promptings of our other instincts was the origin of much human mental conflict. Our herd instinct, moreover, was the key to understanding why "a very considerable proportion" of human beliefs were nonrational.[22] Trotter contended that although most individuals would argue that their beliefs were rationally founded, a close examination would show that "on the great majority of these questions there could be . . . but one attitude—that of suspended judgment."[23] What was accepted as rational foundation, in fact, was merely rationalization.

Tansley agreed with Trotter that the tendency to rationalize was strong in man. He also accepted Trotter's contention that the new Freudian perspective suggested the possibility of an objective analysis that could go beyond the rule of the herd, that is, the customs of the day, or what were called "moral codes." A reflective mind

22. Wilfred Trotter, Instincts of the Herd in Peace and War, 1916–1919 (London: T. Fisher Unwin, 1923): 35. Trotter's book was first published in 1916. A second edition came out in 1919. The 1923 edition is a reprint of the second edition. It was reprinted in 1975 by Gale Research Company, Detroit.

23. Ibid., 36.

could free itself from the herd instinct and its formalized external codes.[24] Psychology, according to Tansley, not only explained how and why moral codes regulated behavior but also demonstrated the ego's capacity to go beyond them. "This highly-developed self-regarding sentiment, which enables a man thus to escape from the immediate pressure of the herd, is nothing else than the ethical self, which . . . is the highest authority in conduct, and the existence of which is necessary to moral freedom."[25] The ethical self transcended the herd. Its development was based on the rational judgments of a well-consolidated ego with an appreciation of and respect for its relationship to the herd.[26]

Tansley and others saw modern psychology providing a new analysis of the feeling of moral obligation. For them, its scientific insights surpassed the vague writings in British moral philosophy on human sympathy and established a suitable foundation for understanding morality. However, it is not clear that it actually solved any of the philosophical issues engulfing evolutionary ethics. Trotter and Tansley exemplified an attempt to understand the psychic origins of our altruistic promptings. Our biological background, according to them, predisposed us to conform to the conduct of our peers and to feel a sense of obligation to them. But such a sense, they admitted, was merely a development that we shared with wolves and sheep. "We must be free from the unchecked sway both of our instincts and of all external moral codes before we can set reason free to do its work in the ethical sphere," Tansley wrote.[27] But did not this recognition of the independence of the moral domain vitiate any help that psychology could offer philosophy? Sidgwick, among others, had argued that the rational examination of morals was independent of their origins or psychological motivation.

As the history of evolutionary ethics demonstrates, however, scientists sometimes have rushed in where philosophers fear to tread. In spite of the clear separation that many psychologists made between the higher ethical self and the deeper psychological currents of the mind, a new factor had been made available to the

24. A. G. Tansley, The New Psychology and Its Relation to Life (London: George Allen and Unwin, 1920): 176.
25. Ibid., 189.
26. Ibid., 189–190.
27. Ibid., 176.

discussions on evolutionary ethics. And just as psychoanalysts insisted that the rational mind could take the confused and self-rationalizing individual ego to a healthier, better-integrated, more honest self, so, too, did several biologists believe that the Freudian revolution could complement the Darwinian one and lead to a revitalized evolutionary ethics. Julian Sorell Huxley and C. H. Waddington were the most well known figures in the life sciences who took this position.

Champions of Evolutionary Ethics

Julian Huxley

Julian Huxley was born in the house of his aunt, Mrs. Humphry Ward, and the connection was significant. Her best-selling novel, Robert Elsmere (1888), resolved its hero's crisis of conscience by applying his religious energy to secular problems. The book made a deep impression on Huxley, and in his memoirs he indicated, "[It] helped convert me to what I must call a religious humanism, but without belief in any personal God."[28] Like his famous grandfather, Julian Huxley was locked in an ambivalent battle with organized religion all his life and saw in science a new force to guide action. But unlike Thomas Henry Huxley, Julian Huxley believed that science could not only tell us how to do things but that it could also supply a foundation for discussing what we ought to do.

Julian Huxley was one of the architects of the modern synthesis, the neo-Darwinian theory that reasserted Darwin's emphasis on the natural selection of small random variations as the central driving force of evolution. His importance derived more from his ability to synthesize the information and ideas that were current at the time than in formulating new biological concepts. Like his grandfather, who also did solid but not revolutionary zoological research, his fame rested on his ability to take the ideas of others to a wider audience.

But it was more than just a revitalized evolutionary theory that Huxley espoused. It was also a vision of progressive evolution with significant social and philosophical implications. Huxley be-

28. Julian Huxley, Memories (London: George Allen and Unwin, 1970): 153.

lieved evolution was a process with three distinct phases: physical, biological, and psychosocial. Although he updated the conventional evolutionary picture of reality with the most modern biological information, the general picture was one that was common from Darwin's and Spencer's day: a cosmological evolution stressing the formation of the solar system; a biological evolution from simple organic soup to the pinnacle of biological existence, man; and a cultural evolution from barbarism to civilization and the hope of future higher development.[29]

During the interwar years and their aftermath, intellectuals like Huxley were confronted with an unsettling array of new cults, which were competing to replace the declining accepted creeds. Like their Victorian grandfathers, Huxley's generation sought new foundations for belief and for social organization. For Huxley, the theory of evolution provided a key element for a new humanist faith based on a scientific worldview and a liberal social philosophy.[30] What distinguished Huxley from other humanists was his emphasis on the modern synthesis and his belief that the human phenomenon had to be viewed from a cosmic perspective.

All humanists did not see the necessity of such an inclusive vision. Walter Lippmann, in his Preface to Morals (1929), sketched a moral philosophy based simply on liberal values and modern psychology. For him, the path was adequate, indeed obvious, without the appeal to evolution. "When men can no longer be theists, they must, if they are civilized, become humanists."[31]

But Huxley sought a more "religious humanism," and he believed the key lay in linking his broad social vision with the theory of evolution. If Lippmann had the good sense to realize that all gospels of science "do violence to the integrity of scientific thought and they cannot satisfy the layman's need to believe,"[32] Huxley still

29. One of Huxley's best statements of his evolutionary picture is in his New Bottles for New Wine (London: Chatto and Windus, 1957).

30. Evolutionary humanism was a theme that ran through Huxley's entire career. For a good introduction see Julian Huxley, ed., The Humanist Frame: The Modern Humanist Vision of Life (New York: Harper and Brothers, 1961). Huxley's humanism was part of the broader, twentieth-century humanist movement; see A. J. Ayer, ed., The Humanist Outlook (London: Pemberton, 1968); Paul Kurtz, The Humanist Alternative: Some Definitions of Humanism (London: Pemberton, 1973); and Morris Storer, ed., Humanist Ethics: Dialogue on Basics (Buffalo: Prometheus Books, 1980).

31. Walter Lippmann, A Preface to Morals (New York: Macmillan, 1929): 137.

32. Ibid., 125.

hoped for a vision that would succeed where others had failed. He sought a "religion without revelation" that would overcome Lippmann's caveat on "the difficulty of reconciling the human desire for a certain kind of universe with a method of explaining the world which is absolutely neutral in its intention."[33]

Huxley attempted to establish the reality of biological progress as a background for a broader philosophy. His concluding chapter of Evolution: The Modern Synthesis was an extended argument for the acceptance of evolutionary progress.[34] In it he reviewed the history of life and argued that it reflected a series of dominant types representing advances in increased control of and independence from the environment.[35]

Huxley's general concept of evolutionary progress met with considerable criticism. Other evolutionists of the modern synthesis were uncomfortable with his discussion, even those who were generally predisposed to see some form of progress over time. The paleontologist George Gaylord Simpson, for example, who wrote more about evolutionary trends than any of the other architects of the modern synthesis, was skeptical about the concept of "improvement" that Huxley used to characterize evolutionary change and that served as a link to his ethical ideas.[36]

Huxley's views on ethics were integrated with his writings on evolutionary progress.[37] However, his evolutionary ethics was based

33. Ibid., 131. Huxley's fullest discussion on religious humanism is in his Religion without Revelation (London: Ernest Benn, 1927).

34. This was not a new theme for Huxley. He had argued for it since early in his career. See, for example, his Essays of a Biologist (London: Chatto and Windus, 1923), where he developed most of the ideas on progress that he later published in Evolution: The Modern Synthesis (New York: Harper and Brothers, 1943).

35. Huxley, Evolution, 556–578.

36. George Gaylord Simpson, Biology and Man (New York: Harcourt, Brace and World, 1969): 141. For some recent views see Matthew H. Nitecki, ed., Evolutionary Progress (Chicago: University of Chicago Press, 1988). Stephen Jay Gould also has argued against "distorting the evolutionary record" by squeezing our interpretations of it into progressive molds; see, for example, his Wonderful Life: The Burgess Shale and the Nature of History (New York: W. W. Norton, 1989), and his article "On Replacing the Idea of Progress with an Operational Notion of Directionality," in Nitecki, ed., Evolutionary Progress, 319–336.

37. On Huxley's worldview based on evolutionary progress, see John C. Greene's essay, "From Huxley to Huxley: Transformations in the Darwinian Credo," in his Science, Ideology, and Worldview, and his article "The Interaction of Science and World View in Sir Julian Huxley's Evolutionary Biology," Journal of the History of Biology 23, no. 1 (1990): 39–55. Also see Sister Carol Marie Wildt, "Julian Huxley's Conception of Evolutionary Progress," Ph.D. dissertation, Saint Louis University, 1973.

more on an acceptance of human progress than on the general idea of biological progress. Although he wrote extensively of the evolutionary unity of the world, he nonetheless made a clear distinction between human and biological progress, and for that reason one might accept his arguments for one while holding reservations about the other.

Human progress differed from biological evolution, according to Huxley, in two significant ways. First, it involved only one species, and more important, it was a progress that was not based on any new biological trend. Instead, human progress was predicated on cultural advance. Cumulative transmission of experience had given humans the ability to communicate knowledge beyond one generation. This ability allowed for cultural evolution, which proceeded at a rate much faster than biological evolution. Huxley held that not only was psychosocial evolution faster than biological evolution but also that it was the only line open to future progress. Life had exhausted its potential for major physiological improvement, and human cultural evolution was unique in escaping the dead end of 700 million years of life on earth. Huxley revealed his anthropocentric bias in a curious comment elaborating on this position, where he argued not only that man was the only organism capable of major evolutionary advance but that "even should the conclusion prove unjustified that purely biological evolution has reached its limit and become stabilized, and some new animal type should arise which threatened man's dominant position, man would assuredly be able to discern and counter the threat in its early stages."[38]

Of greater importance for the issue of ethics were the different criteria that Huxley employed in gauging cultural evolution. Unlike other species, humans were unique in their ability to utilize conceptual thought and speech, and those characteristics distinguished humans from all other animals.[39] Or, as he put it in 1941, "There is but one path of unlimited progress through the evolutionary maze."[40] That path was man's mental self-control and mental independence. Intimately connected to that control and indepen-

---

38. Huxley, New Bottles for New Wine, 48. Presumably Huxley meant that a life form that was physically more fit would be destroyed by humans who had the added advantage of culture and could recognize the threat.
39. Julian Huxley, The Uniqueness of Man (London: Chatto and Windus, 1941): 115.
40. Ibid., 16.

dence were the fulfillment of human possibilities and the formation of values. Cultural evolution and the progress of mankind were the result of the struggle between ideas and values. Instead of a struggle for physical existence and reproduction that characterized the rest of the living world, human evolution was judged by cultural evolution. Unlike most of the earlier writers on evolutionary ethics, Huxley did not stress the adaptive value of cultural evolution. Although he certainly did not underrate the importance of man's ingenuity in controlling his environment, or the great advances in man's ability to feed, clothe, and shelter himself, it was by the increase of aesthetic, spiritual, and intellectual satisfaction that Huxley judged human evolution. And it was in the formation of values for their own sake that the future of progress was to be realized.

Like his grandfather, Julian Huxley envisioned a discontinuity between animal and human evolution. However, where T. H. Huxley was satisfied to acknowledge a shared moral sentiment that came from the heart, Julian wanted to use the science of his day to grapple with what "from the heart" meant. His Romanes Lecture (1943) was a conscious attempt to resolve the problems left by his grandfather's lecture of fifty years earlier, and he published an edition of both lectures together in 1947.

Julian Huxley concentrated on two issues that he considered basic. The first concerned moral obligation; the second, moral standards. These not only have been the central concerns of ethics as an intellectual discipline but also have been central in the criticism of evolutionary ethics, and Huxley correctly realized that any attempt to establish an evolutionary ethics would have to confront them.

To explain moral obligation, Huxley relied on the new psychology, particularly Freudian analysis of infantile mental development. It was in the universal conflict between the desires of the infant relative to its mother that a primitive mechanism came into being, what the Freudians called the primitive superego but what Huxley called, using a "more non-committal term," the "protoethical mechanism."[41] As the infant, early in its second year of ex-

---

41. Thomas Henry Huxley and Julian Huxley, *Touchstone for Ethics 1893–1943* (New York: Harper and Brothers, 1947): 117. Julian Huxley's Romanes Lecture first appeared as *Evolutionary Ethics* (Oxford: Oxford University Press, 1943).

istence, distinguishes itself from outer reality, it is its mother who "comes to represent the external world, and to mediate its impacts on the child."[42] The mother did so by being its source of gratification but equally its source of frustrating "authority." The conflict of aggression and love thus engendered was potentially chaotic, but it constituted the proto-ethical mechanism that repressed aggression by branding it with guilt and therefore allowed the infant to act in the face of conflict. This sense of guilt, of basic wrong as opposed to right, was at the heart of the moral sense.

> Our modern knowledge also helps us to understand the absolute, categorical, and other-worldly quality of moral obligation, on which moral philosophers lay such stress. It is due in the first instance to the compulsive all-or-nothing mechanism by which the primitive super-ego operates. It is also due to the fact that, as Waddington points out, the external world first intrudes itself into the baby's magic solipsism in the form of the parents' demands for control over primitive impulses, so that infantile ethics embody the shock of the child's discovery of a world outside itself and unamenable to its wishes.[43]

The primitive moral sense, of course, represented no more than an "embryonic mental structure." Like Lippmann, whom he admired and quoted in spite of their differences, Huxley held that from this early mental structure the main lines of development were yet to come in the "passage to maturity." Individual human mental development could lead to an "excess load of unrealistic guilt" with all the distortions the Freudians described. But in favorable circumstances, "human beings are able to develop without these overdoses of untruth and unreality in their moral system."[44] Humans could "achieve an internal ethical realism," that is, "the proper adjustment of the sense of guilt to reality."[45]

But ethics involved more than explaining our sense of individual moral obligation through psychology, and Huxley discussed it from a social perspective as well. The ethical standards of the social group in which a person could find himself might be "unrealistic." Just as our individual conscience was not an absolute authority, so, too, were social ethics relative to time and place. But

42. Huxley and Huxley, Touchstone for Ethics, 117.
43. Ibid., 120.
44. Ibid., 124–125.
45. Ibid., 125.

were there independent ethical standards? Certainly not in the sense of a set of absolute values, either current or, as in Spencer's judgment, defined by a future society. Instead standards have evolved in the course of human history and could now provide guidance.

> In the broadest possible terms evolutionary ethics must be based on a combination of a few main principles: that it is right to realize ever new possibilities in evolution, notably those which are valued for their own sake; that it is right both to respect human individuality and to encourage its fullest development; that it is right to construct a mechanism for further social evolution which shall satisfy these prior conditions as fully, efficiently, and as rapidly as possible.[46]

Huxley expanded on some of the implications of his position.

> When we look at evolution as a whole, we find, among the many directions which it has taken, one which is characterized by introducing the evolving world-stuff to progressively higher levels of organization and so to new possibilities of being, action, and experience. This direction has culminated in the attainment of a state where the world-stuff (now moulded into human shape) finds that it experiences some of the new possibilities as having value in or for themselves; and further that among these it assigns higher and lower degrees of value, the higher values being those which are more intrinsically or more permanently satisfying, or involve a greater degree of perfection.[47]

The direction of progress, then, was toward human fulfillment and the realization of things that humans judge to have value. The highest of these values were ones of "intrinsic worth." Society, therefore, should be structured so as to promote them. Huxley elaborated on what sort of society he believed would best realize these goals. It was one that respected the rights of its individuals, one that did not warp the structure of individual personalities, one that stressed education, fostered responsibility, encouraged the arts; that is, the liberal society that fellow humanists and progressive Christians envisioned.

Huxley's evolutionary ethics is interesting because it recognized the degree to which culture had gone beyond purely biological development. As a vision of reality, it must be considered on a par with other inspired dreams of reason of our century, like Teilhard

46. Ibid., 136.
47. Ibid., 136–137.

de Chardin's. Huxley's outpouring of essays, which rivaled his grandfather's in bulk, were read eagerly by many who appreciated the richness of modern evolutionary biology and who were searching for a synthesis that preserved traditional, Western liberal values. Huxley's scientific humanism attempted to liberate modern man from outmoded creeds and to be more complete than Marxism, existentialism, or liberal theology.

But it is not clear that he made any progress in resolving the numerous criticisms raised by Sidgwick and others to earlier systems of evolutionary ethics. Indeed, by tying his ethics to a cosmic scheme of progress, Huxley, like Spencer, risked having his argument dismissed without a fair hearing by those who regarded such speculative visions with disdain.

Critics of evolutionary ethics have traditionally focused on its lack of an adequate explanation of moral obligation, on both the individual and the group level. Huxley's move to a psychoanalytical account of moral obligation was similar to the eighteenth- and nineteenth-century attempts to derive our notions of the moral sentiment from psychology. The problem with such strategies, as Sidgwick pointed out, was not the validity of the hypothesis but its relevance for ethics. Even Huxley admitted that the sense of moral obligation was no more than a psychological factor, which, to an important degree, needed to be mediated by the rational mind. Huxley noted that guilt, which was central to the Freudian interpretation of moral obligation, could be psychologically and morally damaging. For example, in his description of the Nazi mind, Huxley claimed that "an excess load of unrealistic guilt" led individuals to project their sense of unbearable condemnation onto Jewish scapegoats. "Most Nazis genuinely believe that Jews are a major source of evil; they can do so because they have projected the beastliness in their own souls into them. The terrible feature of such projection is that it can turn one's vices into virtues: thus, granted the Nazi believes the Jews are evil, it is his moral duty to indulge his repressed aggression in cruelty and violence towards them."[48]

The value of psychoanalysis lay in its ability to probe and resolve rationally the pathological mental state of individuals. How did Freudian explanations illuminate the ethical problem of moral

48. Ibid., 123.

obligation? In Huxley's discussion, they provided an explanation of why humans might feel obligation, but they did not provide an ethical justification for those feelings. The new psychology, therefore, did not supply a solution to the problem Thomas Henry Huxley recognized in evolutionary accounts of the moral sentiment. Like earlier treatments of the moral sentiment that relied on "custom" for answering the philosophical question of why we "ought" to help our neighbor, Julian Huxley's extended psychological derivation failed to justify what it described.

Huxley's discussion of general ethical standards was no more successful. He argued for those actions, policies, and beliefs that furthered human progress. He measured human progress by the realization of values that were intrinsically worthy, the furthering of individual fulfillment, and the changing of society to promote social evolution. But these were very vague categories. What did intrinsically valuable mean? For Huxley they were obvious: aesthetic, spiritual, and intellectual experiences. But aside from asserting their importance, Huxley did not attempt any justification of them. He was clear that they did not contribute to survival. Rather, they were the values of "high culture"; of considerable importance to a small fraction of the population and traditionally the yardstick by which they measured civilization. But they could, and have, been viewed differently: as the tools of repression, the products of vanity, commodities, and so on. Jerry Falwell, Jacques Derrida, the Ayatollah Khomeini, Allan Bloom, Václav Havel, Madonna, and Noam Chomsky (to name just a few) probably would have differed with Julian Huxley about the significance of high culture as well as what was "intrinsically valuable."

Similarly, the value of individual fulfillment and the social engineering to bring it about were values typifying Western, democratic, liberal thought. In Huxley's day, as in ours, it was not universally accepted, even in the West. And even if it were, the issue in ethics has not been consensus but justification.

Huxley's ethics was a projection of his values onto the history of man. In classic Whig fashion, he conceptualized the past as leading to what he valued in the present and posited the direction of future evolutionary progress from the same perspective. Sincere—and often inspired—as his rhetoric was, it ultimately depended on a shared commitment by his reader rather than logical arguments. Cosmic evolution and biological evolution were blind processes, as Huxley described in numerous publications. And yet he demanded his reader accept the idea that this neutral universe had a direction and that man had an obligation (real, not merely psychological) to further its progress. He could do this by creating values and purpose, according to Huxley. But if man created these values and goals, how were we to be assured of their moral value? To project man's beliefs (the ones of which Huxley approved) onto the cosmos was to create a modern myth, and perhaps there is the key to Huxley's position.

From early in his career Huxley wanted to replace the Christian worldview with a scientific humanism. Huxley believed that the difference between his philosophy and religion was that religion was based on the hypothesis of the existence of a god, whereas his views were based on the scientific method. But what he actually proposed was the creation of a new myth, one dressed in the guise of evolutionary biology but nonetheless with an entire set of assumptions, values, and beliefs. Like many of the attempts since Comte to use science to go beyond religion but still maintain the sense of religious awe, Huxley's naturalism assumed the vision he pretended to discover.

C. H. Waddington

C. H. Waddington was another figure who championed evolutionary ethics in the period between the end of the First World War and the creation of the field of sociobiology. Waddington's ideas were developed in conjunction with Huxley, and therefore it is not surprising that there was considerable similarity in their mature views.

Waddington appears to have been more sensitive than Huxley to the contemporary problems that faced the construction of an evolutionary ethics, indeed, any ethical system. In the beginning of his Science and Ethics (1942), he indicated what a leading philosopher had once said to him on the subject: " 'This,' Wittgenstein once said to me, 'is a terrible business—just terrible! You can at best stammer when you talk of it.' "[49] And although Waddington additionally noted that psychoanalysis, anthropology, Marxism, and logical positivism had undermined traditional philosophical justifi-

---

49. C. H. Waddington, ed., Science and Ethics (London: George Allen and Unwin, 1942): 7.

cations of ethics, he still thought that evolution provided a new possibility. For him, "the framework within which one can carry on a rational discussion of different systems of ethics, and make comparisons of their various merits and demerits, is to be found in a consideration of animal and human evolution."[50]

Like Huxley, Waddington held that ethics had been led down blind alleys by philosophers and theologians and that biologists, conscious of the evolutionary history of man, could set the discussion right. The main issues that he attacked were the familiar ones: the origin of moral obligation and the evaluation of ethical standards. His most complete explication of evolutionary ethics was published in 1960 in a book entitled The Ethical Animal. In it he elaborated on earlier writings and attempted to answer critics.

Waddington relied on psychology to explain the origin of our moral sense. He claimed that the human infant was born with an "innate capacity to acquire ethical beliefs but without any specific beliefs in particular."[51] In 1960 when Waddington published The Ethical Animal, theories of infant development had ramified into a wide range of opinion. He acknowledged that the story was not as simple as he and Huxley had envisioned it to be in the 1940s. He nonetheless repeated his affirmation that the infant's interaction with its environment resulted in an "ethicizing creature," that is, one that was accepting of authority.

Waddington thought that the individual also acquired the ability to examine his own ethical beliefs in a rational manner, and this had important evolutionary consequences in deciding between alternative systems of belief. But what were the "supra-criteria of wisdom," as he phrased it? Waddington agreed with Huxley that the criterion of ethical good was that which furthered human evolution, but he attempted to avoid what he took to be Huxley's circular reasoning in attempting to establish the direction of progress. Instead he stressed that we needed to evaluate particular ethical beliefs against the background of evolutionary advance. The question was "how effective it is in mediating this empirically ascertained course of evolutionary change."[52] Waddington used the

50. C. H. Waddington, The Ethical Animal (London: George Allen and Unwin, 1960): 23.
51. Ibid., 26.
52. Ibid., 30.

analogy of evaluating the activity of eating. The newborn infant developed "into a creature that goes in for eating" in a manner that was parallel (even if more dependent on innate factors) to the infant's development into an "ethicizing being."[53] Next the child developed specific food habits, analogous to the formation of specific ethical beliefs. To criticize those eating habits, Waddington directed us to inquire into the function of eating, which was the possible healthy growth of the body. Having determined the function of eating, one could then ask of a particular eating habit "how effective it is in bringing about healthy growth."[54] If an individual asserted that he preferred abnormal growth, the nutritionist could tell him that to do so would be out of step with nature. The nutritionist was not imposing values but was revealing "biological wisdom" based on what was "immanent in nature." That is, there was an empirical grounding for the assertion. Extending this to ethics, one had to ask of a particular belief, did it contribute to human evolution? By human evolution, Waddington did not mean to imply a simplistic notion but rather "to refer to all the cultural changes which differentiate human life at the present day from that of our Stone Age ancestors. It includes spiritual and intellectual changes as well as those concerning materials and tools."[55] To the criticism that his criterion was so broad as to be meaningless, Waddington replied, "No general ethical principle can be useful unless it is wide enough to be relevant to very many diverse aspects of life; and that implies that it cannot be precise enough to obviate the need for debate about particular moral issues."[56]

That ethical principles must be broad no one would dispute. General principles cannot be expected to deal up directly solutions to particular moral questions. The "evolution of man," however, is so vague and potentially contains so many conflicting empirical examples that it is difficult to see how it, in itself, can be of any use as a criterion for ethical value. Nations have come and gone. War and peace have reigned. There are no guideposts along the way that say, "This is the right direction." The destruction of the Roman empire was of value to some of its neighboring tribes and had many profound effects on Western civilization; some positive,

53. Ibid.
54. Ibid.
55. Ibid., 31.
56. Ibid.

some negative. Was the event good? Waddington criticized Huxley for the circularity of his thought, but it is not clear that he avoided the same pitfall. Where Huxley claimed that emergent values were obvious, Waddington said they could be empirically discovered by reference to human progress. But for both, what ultimately mattered were their own values, which they read into the human drama and which remained philosophically unjustified.

Biologists on Evolutionary Ethics

Although Huxley and Waddington were the two most strident voices relating evolution to ethics, they were by no means the only ones in the interwar and postwar period. Indeed, many evolutionary biologists, or scientists who studied related subjects like genetics and ecology, expressed ideas on the relationship of evolution to moral and/or religious ideas. Most were critical of the attempt to develop an evolutionary ethics along the lines laid out by Huxley and Waddington, but certainly not all. Several biologists in the United States favored a naturalistic theory of ethics that was tied to their ideas on evolution. Animal ecologists at the University of Chicago, for example, based their ethical views on an alternative evolutionary perspective that emphasized the importance of cooperation over natural selection. Building on the cooperationist ideas of earlier writers like William Patten, the Chicago ecologists Warder Clyde Allee and Alfred Edwards Emerson spoke out for a social program and an ethical stance that they justified with their scientific writings on progressive animal evolution.[57] Panels at American Association for the Advancement of Science (AAAS) meetings and popular writings in the 1940s and 1950s also stressed a biological perspective on morality and moral questions.[58] Although evolution was a critical concept in some of these views, the

57. For an interesting discussion see Gregg Mitman, "From Population to Society: The Cooperative Metaphors of W. C. Allee and A. E. Emerson," Journal of the History of Biology 21, no. 2 (1988): 173–194, and his excellent book The State of Nature, which describe the development of animal ecology at the University of Chicago and demonstrate the way in which figures like Allee and Emerson read their social views into the natural world.

58. A good example of this literature is Chauncey Leake and Patrick Romanell, Can We Agree? A Scientist and a Philosopher Argue about Ethics (Austin: University of Texas Press, 1950), which contains Leake's well-known article, "Ethicogenesis," published originally in the Scientific Monthly 60 (1945): 245–253.

ethical position presented was a naturalistic ethics that drew heavily on ethnology, psychology, and physiology and usually lacked the general picture of evolutionary progress that was so central to other biological interpretations of ethics. A well-known example of these naturalistic theories of ethics was Life and Morals (1948) by Samuel Jackson Holmes of the University of California. Holmes claimed that morals had a natural origin but, he said, Darwinian evolution "does not logically compel me to adopt any one standard of conduct rather than another."[59] He accepted Edward Westermarck's relativity of ethics and advocated "human welfare" as the most acceptable standard by which to judge conduct.[60]

There were, then, biologists who expressed a sympathy with those who sought to illuminate the human condition with insights from natural science. But overall, the goal was one that failed to attract a wide following. The major architects of the modern synthesis, with the exception of Julian Huxley, were skeptical of the alleged direct ethical implications of evolution. Theodosius Dobzhansky, who wrote extensively on his philosophical views in which evolution plays an important role, was highly critical of evolutionary ethics for not giving enough attention to the role of free choice. In his Page-Barbour Lectures at the University of Virginia (1954), he cautioned against accepting too simple a view of progress and then asked, "Suppose, however, that future studies of human biology and evolution tell us exactly what the direction of evolution in general, and of human biology in particular, has been. Just why should we take for granted that this direction, which we have not chosen, is good?"[61] He emphasized this weakness in the basic foundations of any evolutionary ethics. "No theory of evolutionary ethics can be acceptable unless it gives a satisfactory explanation of just why the promotion of evolutionary development must be regarded as the summum bonum."[62] Furthermore, in Dostoyevskian fashion, Dobzhansky insisted that man would not deny himself the right to question anything, "including the wisdom

---

59. Samuel Jackson Holmes, Life and Morals (New York: Macmillan, 1948): 41.
60. Edward Westermarck's Origin and Development of Moral Ideas (1906–1908) and his shorter Ethical Relativity (1932) were among the most widely cited books on ethical relativism at this time.
61. These lectures were published as Theodosius Dobzhansky, The Biological Basis of Human Freedom (New York: Columbia University Press, 1956); see 129.
62. Ibid., 128.

of his evolutionary direction."[63] Following out this line of thought, he said that man was free to rebel, that even if the direction of evolution were demonstrated to be "good," "man is likely to prefer to be free rather than to be reasonable."[64]

George Gaylord Simpson, another figure of central importance to the modern synthesis who published his ethical views, was outspoken in his criticism of Huxley's and Waddington's attempts to resuscitate evolutionary ethics. Simpson agreed that the moral sense had a natural, evolutionary origin; perhaps not exactly in the fashion that Huxley or Waddington favored but in some psychosocial manner. However, such an origin did not proscribe any specific ethical conclusions. "A rational, naturalistic system of ethics cannot be independent of evolution, but neither can it be derived from evolution."[65] Like Dobzhansky, Simpson claimed that what makes ethical choice valid is man's freedom to make ethical choices among alternatives. The past evolution of man did not supply a guide, nor did any perceived direction in evolution. Simpson disagreed with Huxley on the nature of biological progress and instead stressed the opportunistic nature of evolution. Far from seeing a straight line from protozoa to man, Simpson believed "evolution is not invariably accompanied by progress, nor does it really seem to be characterized by progress as an essential feature. Progress has occurred within it but is not of its essence. Aside from the broad tendency for the expansion of life, which is also inconstant, there is no sense in which it can be said that evolution is progress."[66] Although the fossil record revealed man to be "high," progress in the evolution of man was relative to him only. "It is not progress in a general or objective sense and does not warrant choice of the line of man's ancestry as the central line of evolution as a whole."[67]

In his widely read book The Meaning of Evolution (1949), Simpson reviewed the various attempts to establish an evolutionary ethics from Darwin's day to the present. He criticized both the

63. Ibid., 129.
64. Ibid.
65. Simpson, Biology and Man, 142.
66. George Gaylord Simpson, The Meaning of Evolution: A Study of the History of Life and of Its Significance for Man (New Haven: Yale University Press, 1949): 261. This book was an expanded version of the Terry Lectures Simpson delivered at Yale in 1948.
67. Ibid., 262.

tooth and claw ethic that Thomas Henry Huxley rejected and the
more positive "life ethics" of writers like Spencer who defined ethical conduct in terms of those actions that promote life. Simpson
claimed that the problem with attempts to derive an evolutionary
ethic had to do partly with the quest for an absolute ethic, and he
concluded that

> the search for an absolute ethic, intuitive or naturalistic, has been a
> failure. Survival, harmony, increase of life, integration of organic or
> social aggregations, or other such suggested ethical standards are char-
> acteristics which may be present in varying degrees, or absent, in or-
> ganic evolution but they are not really ethical principles independent
> and absolute. They become ethical principles only if man chooses to
> make them such. Man cannot evade the responsibility of the choice.[68]

There was, then, no possibility in deriving an ethic from past evolutionary history. Instead, man had to choose to develop his capacities and make ethical decisions. From this position, similar
to Dobzhansky's, Simpson drew closer to Huxley's and Waddington's. For he went on to state that human ethical standards must
be relative to man's evolution. "Evolution has no purpose; man
must supply this for himself. . . . But human choice as to what are
the right ends must be based on human evolution."[69] Simpson took
these basically "good" goals of human conduct to be the promotion of knowledge, the recognition of personal responsibility, the
recognition of the integrity and dignity of individuals, and the importance of promoting the fulfillment of individual capacities.[70]
    Although Simpson sought a naturalistic ethic, one that was
linked to evolution, he was more aware than Huxley or Waddington of its relative nature. Citing Westermarck's anthropological
arguments on the diversity of ethical beliefs and the Freudian revolution in our understanding of our intuitive sense of moral obligation, Simpson was keenly aware of the limits of his position.
Nonetheless, he did conclude that "these relativistic ethics have, at
least, the merit of being honestly derived from what seems to be
demonstrably true and clear."[71]

    68. Ibid., 310.
    69. Ibid., 311.
    70. Ibid., 315. For an interesting discussion of Simpson's emphasis on the in-
dividual in contrast to the group, see Mitman, "From Population to Society,"
190–193.
    71. Ibid., 324.

Considering the diversity of the writings of Morgan, Machin, Huxley, Waddington, Holmes, and Simpson, one might ask, Is there an internal logic that leads those strongly committed to the theory of evolution to attempt an ethics informed by (if not derived from) evolution? Probably not. There have been striking counterexamples; Dobzhansky, to mention one. In addition, there were biologists committed to evolution who approached the topic of ethics from quite different perspectives. For example, Joseph Needham argued that life consisted of more than one irreducible form of experience and that in addition to scientific, there were also philosophical, historical, aesthetic, and religious dimensions to our life.[72] His essays depict a highly individualistic, syncretistic worldview, which partakes of the Tao, Marxism, and Christianity as well as biochemistry and evolution.[73]

But if there is no internal logic that leads from evolution to ethics, there certainly has been a strong invitation to construct an ethics from an evolutionary perspective. The theory of evolution, especially in its modern form, is a brilliant synthesis of biological knowledge and gives direction to much of the most interesting work on life. Since man is a product of evolution, it has been tempting to see if the theory that explained so much that formerly was unexplainable could provide some illumination on this intractable problem. Scientific theories often grow by extension. The field of ethics has been particularly attractive to scientists because in the secular atmosphere of the twentieth century, religion has not provided an accepted foundation for ethical belief. As we have seen, this has been a long-standing problem. Since the crisis of the 1870s, the situation has steadily deteriorated. Moral philosophy in the second half of the nineteenth century attempted to supply a new foundation, either through idealism or utilitarianism, but could not arrive at consensus. The proponents of evolutionary ethics can

72. See Needham's autobiographical sketch written under a pseudonym, Henry Holorenshaw, "The Making of an Honorary Taoist," in Mikuláš Teich and Robert Young, eds., Changing Perspectives in the History of Science: Essays in Honour of Joseph Needham (London: Heinemann, 1973): 5. Also see the seminal book by R. G. Collingwood, Speculum Mentis: Or the Map of Knowledge (Oxford: Oxford University Press, 1924).

73. See Needham's three books of essays, The Sceptical Biologist (London: Chatto and Windus, 1929); The Great Amphibium: Four Lectures on the Position of Religion in a World Dominated by Science (London: Student Christian Movement Press, 1931); and Time: The Refreshing River, Essays and Addresses, 1932–1942 (London: George Allen and Unwin, 1943).

be seen as individuals who, dissatisfied with religious and philosophical treatments of ethics, attempted to supplant them and join
the ranks of those who offered serious philosophical accounts. But
their efforts were judged wanting by professional philosophers in
both the late nineteenth and the early twentieth century. Did the
climate in moral philosophy alter between G. E. Moore's publication of Principia Ethica and the elaboration of the modern synthesis? Even a cursory glance at the literature reveals that philosophers did not soften their antagonism to evolutionary ethics. But
did moral philosophers formulate an acceptable alternative?

Evolutionary Ethics and Twentieth-Century Moral Philosophy

Twentieth-century Anglo-American philosophers have failed to
come to any agreement on the foundations of ethics. G. E. Moore,
who began a new approach to ethics early in the century, argued
that the "good" was not definable. That is, one could not identify
natural properties to characterize objects or actions that were
good. Instead moral truths were known to us by direct intuition.
Moore wrote in his Principia Ethica,

> By far the most valuable things, which we know or can imagine, are
> certain states of consciousness, which may be roughly described as
> the pleasures of human intercourse and the enjoyment of beautiful ob-
> jects. No one, probably, who has asked himself the question, has ever
> doubted that personal affection and the appreciation of what is beauti-
> ful in Art or Nature, are good in themselves; nor, if we consider strictly
> what things are worth having purely for their own sakes, does it ap-
> pear probable that any one will think that anything else has nearly so
> great a value as the things which are included under these two heads.[74]

Personal affections and aesthetic enjoyment, consequently, were
the ends by which we should judge actions. "Our 'duty,' therefore,
can only be defined as that action, which will cause more good to
exist in the Universe than any possible alternative. And what is
'right' or 'morally permissible' only differs from this, as what will
not cause less good than any possible alternative."[75]
Although Moore's philosophy impressed the Bloomsbury crowd

74. Moore, Principia Ethica, 188–189.
75. Ibid., 148.

and a significant number of philosophers, his ethics was contested by many. Among those who accepted the notion that moral truths were known by intuition, important philosophers like H. A. Prichard and W. D. Ross did not agree with Moore's contention that the moral value of an action should be judged by the end toward which it led but held instead that it was intrinsic to the action itself. Anglo-American logical positivists, who came to occupy an important niche in the discipline, however, rejected the notion of intuitive moral knowledge altogether. They did so because all knowledge, according to the logical positivists, had to be verified empirically. Intuitive notions were not considered in the realm of verifiable statements. A. J. Ayer, for example, dismissed the entire field of ethics because he held that moral discourse was beyond empirical verification. "We can now see why it is impossible to find a criterion for determining the validity of ethical judgements," he wrote after examining various ethical positions.

> It is not because they have an "absolute" validity which is mysteriously independent of ordinary sense-experience, but because they have no objective validity whatsoever. If a sentence makes no statement at all, there is obviously no sense in asking whether what it says is true or false. And we have seen that sentences which simply express moral judgements do not say anything. They are pure expressions of feeling and as such do not come under the category of truth and falsehood. They are unverifiable for the same reason as a cry of pain or a word of command is unverifiable—because they do not express genuine propositions.[76]

Indeed, for Ayer, ethics did not belong in the province of philosophy but was "a task for the psychologist."[77] By that he meant that ethical statements were expressions of feeling, not statements that had a truth value.

Logical positivism had a major impact on Anglo-American philosophy and contributed to the notion that ethics had more to do with feeling than knowledge. Starting from this position and emphasizing the importance of language, Charles L. Stevenson elaborated a sophisticated and widely influential approach to the clarification of the concepts of ethical discourse and their use: the

76. A. J. Ayer, Language, Truth and Logic (London: Victor Gollancz, 1936): 161.
77. Ibid., 168.

"emotivist" theory. "Moral judgments are concerned with recommending something for approval or disapproval," he wrote.[78] After the Second World War philosophers in England and America continued to stress the study of language. Many were inspired by John Langshaw Austin, who established a tradition of ordinary language philosophy at Oxford which analyzed the meaning of everyday language. By focusing on how ethical terms are used, by clarifying what we mean by ethical judgments, or by dismissing the cognitive value of ethics altogether, followers of the analytical traditions in the Anglo-American philosophical community have all but abandoned attempts at the construction of a rational foundation for ethics.

And so it is understandable that alternative approaches to ethics would be explored in the first half of the twentieth century. Some of those searching for a new base for ethics looked to biology to supply what philosophy and religion have been unable or unwilling to construct. Because part of biology's domain dealt with human origins and with psychological phenomena, it was a reasonable quest. Unfortunately, biology did not prove to be fertile ground. How could it? As we have seen, a common assumption of writers on human evolution was that with the appearance of man a new phase of evolution came into being: human culture. The key to evolutionary ethics, therefore, was to be found in human cultural evolution, an area poorly understood by biology, if understood at all. For that reason, critics of evolutionary ethics argued that the origins and foundations of ethics were outside the domain of biology and rather belonged in the province of anthropology, history, philosophy, and social science. Attempts at speculative evolutionary philosophy with the aim of providing a basis for moral philosophy have produced little more than projections of value systems onto human evolution. Evolution may have proven useful in influencing conceptions of human nature, but in itself that knowledge was insufficient to generate an ethics.

78. Charles L. Stevenson, Ethics and Language (New Haven: Yale University Press, 1944): 13. Emotivist theory was just one of a number of positions that derive from the study of the language of ethical discourse. R. M. Hare developed an alternative called "prescriptivism" that has been widely discussed and extended. See William Donald Hudson, Modern Moral Philosophy, 2d ed. (London: Macmillan, 1983), and Douglas Seanor and N. Fotion, eds., Hare and Critics: Essays on "Moral Thinking" (Oxford: Oxford University Press, 1988).

A review of evolutionary ethics since the First World War, then, shows that it had a second chance to die a natural death. But Phoenix-like, it reappeared in the 1970s. In part, the continued dissatisfaction with philosophy's inability to provide an agreed upon moral philosophy provided an incentive to continue to explore biological alternatives. But of greater importance was the dramatic development of sociobiology in the later part of the twentieth century, which led numerous writers to believe that science had achieved a breakthrough that would illuminate the secrets of human nature and open the possibility (finally) of constructing a valid evolutionary ethics.

8. Evolutionary Ethics Since 1975

Dodo, Phoenix, or Firebird?

The first episode in the history of evolutionary ethics began during the nineteenth century with Darwin's and Spencer's writings. Although university dons and professors were skeptical about the relevance of this new perspective on morality, numerous scientists and writers were enthusiastic. By the beginning of the twentieth century, however, the verdict of informed opinion was negative, even though the topic had not been banished from respectable, educated discourse or from university curricula. Until the mid-1930s, the Moral Sciences Tripos at Cambridge contained such questions as "Discuss the position occupied by Evolution in Mr Spencer's system of Ethics," and "What are the principal ethical problems upon which the theory of Evolution has been supposed to throw light? Consider how far it really does throw light upon them."[1] But by the end of the First World War, the subject had encountered severe criticism and was discussed less frequently in the learned journals. Nonetheless, it had a popular appeal and reached a broad audience of general readers.

Attempts to resurrect evolutionary ethics as a subject for serious philosophical consideration in the first half of the twentieth century constituted the second episode in its history. The most thoughtful of the new contenders were biologists like Julian Huxley or C. H. Waddington who were operating with a newly formulated theory of evolution and who believed Freudian psychology supplied additional support. Their efforts, however, failed to generate widespread acceptance. Professional philosophers were dis-

1. Cambridge University Examination Papers (Cambridge: Cambridge University Press, 1899–1987). The first question quoted is from vol. 30 (1900–1901): 359; the second question is from vol. 39 (1909–1910): 430. After vol. 66 (1936–1937), questions on evolutionary ethics do not appear in the Tripos with the single exception of vol. 79 (1949–1950): 608: "Discuss the view that a line of action is right if and only if it is in harmony with the trend of evolution."

148

dainful, and for the most part evolutionary ethics continued to be confined to popular literature.

Some scientists were sympathetic, but in the highly specialized intellectual environment of the twentieth century, their "amateur" writings had little impact. Like Samuel Jackson Holmes, these writers were concerned to demonstrate general issues such as the natural origin of the moral sense. Symposia at scientific meetings occasionally focused on these subjects. At one meeting of the AAAS, for example, the History and Philosophy of Science Section sponsored a panel on "Science and Ethics" at which it was agreed that biological generalizations have moral consequences. The "importance of harmonious adaptation" was typical of the "biological generalizations" thought to have ethical import, and the AAAS group proposed that "the probability of survival of a relationship between individual humans or groups of humans increases with the extent to which that relationship is mutually satisfying and advantageous."[2] This bromide was made more "rigorous" and was restated: "The probability of survival of individuals, groups, or species of living things increases with the degree with which they can and do adjust themselves harmoniously to each other and to their environment."[3] Social customs, according to this point of view, had "exhibited survival value in a Darwinian sense."[4]

These forays into philosophy elicited little positive reaction. Perhaps the lack of a new major monograph on evolutionary ethics partly explained the absence of discussion. Sporadic symposia and occasional pieces were not enough to kindle a serious debate. Huxley's Romanes Lecture was a slim work, and Waddington's Science and Ethics was only an essay and a set of responses. His more extensive Ethical Animal did not appear until 1960, by which time evolutionary ethics was beginning to take on the characteristics of an intellectual dodo bird.

Edward O. Wilson's exciting synthesis of animal behavior and evolution completely altered this situation and supplied an impetus for a reappraisal of evolutionary ethics, thus opening up a third episode in the history of evolutionary ethics that continues to the present. Wilson's book Sociobiology (1975) gives a name to

2. Leake and Romanell, Can We Agree? 25.
3. Chauncey Leake, "An American Opinion," in Waddington, Science and Ethics, 133.
4. Ibid.

this approach in the study of behavior which stresses the genetic components of behavior and relates them to their evolutionary significance. "Sociobiology" had been studied by biologists before Wilson published his book, and even the term had been used, but Wilson's classic 700-page treatise was what brought the field wide recognition and perhaps more publicity than desired.[5] The final chapter of Sociobiology considers the behavior of man. It raised a storm among those who saw objectionable political and social implications in Wilson's stress on the biological bias of human behavior. Emphasis on human hereditary traits in understanding human nature carries with it an accumulated historical baggage of uncritical, often irresponsible, ideas on genetic determinism, eugenics, and race hygiene. Charges of sexism, racism, and exploitative capitalism were hurled at Wilson, and the 1978 AAAS meeting was disrupted as hecklers took over the stage at a session where he was to speak, a scandal that rivaled the famous ones of the AAAS's British cousin.[6]

A relatively short section, barely two pages, in Wilson's notorious final chapter of Sociobiology discusses ethics.[7] He comes immediately to his point: "Scientists and humanists should consider together the possibility that the time has come for ethics to be removed temporarily from the hands of the philosophers and biologicized."[8] His larger goal, raised in the introduction of the book, is that sociobiology, by explaining the biological basis of social behavior, might become the instrument through which the foundations of the social sciences and the humanities could be reformulated and integrated into the Darwinian worldview.[9] For ethics, this means understanding its "genetic evolution." As

5. The term "sociobiology" was first used by John Scott in 1946. See Edward O. Wilson, "Introduction: What Is Sociobiology?" in Michael Gregory, Anita Silvers, and Diane Sutch, eds., Sociobiology and Human Nature (San Francisco: Jossey-Bass, 1978): 3.
6. Some of the initial intellectual debate can be found in Arthur Caplan, ed., The Sociobiology Debate: Readings on Ethical and Scientific Issues (New York: Harper and Row, 1978). See Degler, In Search of Human Nature, for the background for discussions on human nature in this century; and Ullica Segerstrale, "Colleagues in Conflict: An 'In Vivo' Analysis of the Sociobiology Controversy," in Biology and Philosophy 1, no. 1 (1986): 53–87, for a sociological analysis of the controversy.
7. Edward O. Wilson, Sociobiology: The New Synthesis (Cambridge: Harvard University Press, 1975): 562–564.
8. Ibid.
9. Ibid., 4.

Wilson explains, "Ethical philosophers intuit the deontological canons of morality by consulting the emotive centers of their own hypothalamic-limbic system. . . . Only by interpreting the activity of the emotive centers as a biological adaptation can the meaning of the canons be deciphered."[10] Or, in simpler language, biology can explain in evolutionary terms those moral intuitions that philosophers have so desperately tried to justify.

Although Wilson's specialty is the social insects, he has considerable interest in the scientific explanation of man and his culture.[11] Like Darwin, he sees that a complete evolutionary picture of the world requires an account of man's cultural evolution, specifically, his moral sense. Whereas Darwin was working in a period of hopelessly inadequate genetics, no population biology to speak of, and only the crudest data on ethology, ecology, and psychology, Wilson has the fruits of this century's spectacular progress in these areas. He also shares with numerous other intellectuals a common sense of frustration over the Anglo-American philosophical community's failure to illuminate the rational foundation of ethics. His other motives, whether crypto-sexist, racist, vegetarian, or whatever, need not concern us. Wilson's interest in a complete Darwinian worldview is clear, and he has devoted three books to the topic since publishing his magisterial synthesis on sociobiology.[12]

In these books some of the key concepts of the modern study of the genetic basis of social behavior are popularized. Wilson argues for an understanding of culture from a biological perspective. Central to his discussion is the nature of altruism and the evolution of culture. The two subjects are linked through a study of genetics. A major topic for ethologists has been altruistic behavior, which is defined as behavior that increases the fitness of another at the expense of the actor's fitness. Fitness is understood as genetic fitness,

10. Ibid., 563.
11. Wilson is one of the most creative and wide-ranging biologists alive today. He has made contributions to population biology, written very intelligently on the future of biological research, and is a leading spokesman on the importance of biodiversity.
12. Edward O. Wilson, On Human Nature (Cambridge: Harvard University Press, 1978); Charles J. Lumsden and Edward O. Wilson, Genes, Mind, and Culture: The Coevolutionary Process (Cambridge: Harvard University Press, 1981); and Charles J. Lumsden and Edward O. Wilson, Promethean Fire: Reflections on the Origin of Mind (Cambridge: Harvard University Press, 1983). On the broader context of Wilson's ideas, see Segerstrale, "Colleagues in Conflict."

that is, the relative genetic contribution of one genotype to the next generation compared to other genotypes. That an animal should sacrifice its life for the sake of its neighbors has been a puzzle for over a century. Darwin and his contemporaries relied on the concept of group selection. A group that harbored such individuals would have an advantage over those groups that did not. In recent years, however, the concept of group selection has been strongly criticized. Assaults by George C. Williams, Richard Dawkins, and William Hamilton have nearly destroyed the notion, although some still give the concept a limited value.[13] Sociobiology, for the most part, has emphasized individual selection and builds on the work of biologists who have emphasized the central importance of an individual passing on its genes. For example, through Hamilton's concept of kin selection or "inclusive fitness," the selective value of an altruistic act can be explained by showing how the act results in a greater number of an individual's genes being passed on to the next generation. This can happen, even at the expense of an individual's life or fertility, when the "altruistic" act leads to the survival and reproduction of near relations with whom he shares common genes. Since we share half our genes with siblings (an eighth with cousins), if we sacrifice ourselves so that one of our siblings more than doubles his reproductive rate, copies of our genes in the next generation will be increased. A related concept, "reciprocal altruism," introduced by Robert Trivers, accounts for those cases involving individuals not closely related by coupling "unselfish" behavior to the likelihood of future benefit to the good samaritan.[14]

To be sure, the use of terms like "selfish," "altruistic," and "good samaritan" is misleading, for it suggests conscious intention, choice,

13. See William Hamilton, "The Genetical Theory of Social Behaviour," 2 pts., *Journal of Theoretical Biology* 7 (1964): 1–16, 17–52; George C. Williams, *Adaptation and Natural Selection: A Critique of Some Current Evolutionary Thought* (Princeton: Princeton University Press, 1966); and Richard Dawkins, *The Selfish Gene* (Oxford: Oxford University Press, 1976). The concept of group selection has had a long and tangled history. The modern critique was aimed at the formulation of V. C. Wynne-Edwards who, in his 1962 *Animal Dispersion in Relation to Social Behaviour* (Edinburgh: Oliver and Boyd, 1962), attempted to explain the regulation of population size by hypothesizing that individuals in Mendelian populations sacrifice their survival and limit their reproduction to help control population growth.

14. Robert Trivers, "The Evolution of Reciprocal Altruism," *Quarterly Review of Biology* 46, no. 4 (1971): 35–75. Suggestive as this concept has been, many biologists doubt the extent of reciprocal altruism in nature.

and human values. Georg Breuer, in a very perceptive discussion of the debate surrounding sociobiology, cites the sloppy use of language as partly responsible for some of the many misunderstandings that bedevil the subject.[15] But language is not the sole problem. If we accept the idea that human populations evolved like other social organisms, a clear invitation to consider human behavior in evolutionary terms exists. Might not concepts that shed light on the altruistic behavior in animals be used to understand human altruism also?

But that raises the issue of the origin of culture, for discussions of altruism, values, and kinship relations have to be related to the cultural milieu in which they occur. Here the topic gets cloudy, for during much of this century "cultural" has been used as a term in conscious contrast with "biological."[16] Like Darwin, however, most sociobiologists maintain that there is a continuity between animals and man and that rudiments of culture are to be found among animals. But, equally, sociobiologists concede that culture is primarily a human property. The key issue is, to what extent is culture genetically determined? Sociobiologists differ in their assessments, and these differences are vital. Wilson published two books on the relationship between culture and genes. He argues, along with his collaborator Charles Lumsden, that although a genetic component in humans is responsible for behavioral "tendencies," cultural evolution has to be seen as a coevolution of genes and culture. Culture has units that can be "inherited" in a manner analogous to genes but by a process that is faster, more directed, and more flexible.

Richard Dawkins, a leading popularizer of the evolutionary perspective on societies, animal and human, agrees that concepts like reciprocal altruism and inclusive fitness are inadequate to explain fully cultural evolution. He also accepts the idea that cultural units called "memes" exist, and that "memes and genes may often rein-

15. Georg Breuer, Sociobiology and the Human Dimension (Cambridge: Cambridge University Press, 1982). Also see the interesting discussion of altruism by David Sloan Wilson, "On the Relationship between Evolutionary and Psychological Definitions of Altruism and Selfishness," Biology and Philosophy 7 (1992): 61–68.

16. See Degler, In Search of Human Nature, for an interesting discussion of the history of the distinction. Also see Hamilton Cravens, The Triumph of Evolution: American Scientists and the Hereditary-Environment Controversy 1900–1941 (Philadelphia: University of Pennsylvania Press, 1978).

force each other, but they sometimes come into opposition."[17] Such an admission, in the tradition of Thomas Henry Huxley, opens the door to a view of culture as a human phenomenon that goes beyond biological evolution. In spite of a general perception that Dawkins espouses a "selfish" vision of man, he ends his book on the selfish gene with a statement of hope and a discussion of man's freedom to rebel against his "selfish" legacy.[18] Considerable debate centers on this point. If culture is independent of biological evolution, then who needs sociobiologists intruding into the domains of sociology, anthropology, history, and philosophy? Human geneticists like Luigi Luca Cavalli-Sforza believe that a theory of culture can be constructed which makes use of analogies to Darwinian evolution and to patterns discovered in epidemiology. And ecologists such as Robert Boyd and Peter Richerson have attempted to construct a Darwinian theory of the evolution of culture.[19] Anthropologists like Marshall Sahlins, however, stridently oppose incursions of sociobiology into anthropology and argue that culture has to be understood on its own terms.[20] More accommodating, Ashley Montagu accepts a genetic basis for much behavior but dismisses the approach of sociobiology as narrow "biologism."[21] If, however, sociobiology can tell us something about human nature, or about human social activity, might it not be relevant for moral discussion? Even philosophers who verge on the intemperate in sociobiology bashing appreciate the value of an expansion of our understanding of human nature. Mary Midgley,

---

17. Dawkins, The Selfish Gene, 213. Also see Juan Delius, "The Nature of Culture," in M. S. Dawkins, T. R. Halliday, and R. Dawkins, eds., The Tinbergen Legacy (London: Chapman and Hall, 1991): 75–99.

18. Mary Midgley, for example, in her article "Gene-Juggling," in Ashley Montagu, ed., Sociobiology Examined (New York: Oxford University Press, 1980): 108–134 (originally in Philosophy 54, no. 210 [1979]: 439–458), has a rather savage discussion in which she refers to Dawkins's position as "slapdash egoism" (132). She argues that Dawkins's central point is that the emotional nature of man is exclusively self-interested and that he attempts to make this point by claiming that all emotional nature is self-interested (109).

19. See L. L. Cavalli-Sforza and M. W. Feldman, Cultural Transmission and Evolution: A Quantitative Approach (Princeton: Princeton University Press, 1981), and Robert Boyd and Peter Richerson, Culture and the Evolutionary Process (Chicago: University of Chicago Press, 1985).

20. See Marshall Sahlins, The Use and Abuse of Biology: An Anthropological Critique of Sociobiology (Ann Arbor: University of Michigan Press, 1976).

21. Montagu, Sociobiology Examined, 5. Also see Degler, In Search of Human Nature, 310–327.

for example, who rakes Dawkins over the coals for his "fatalism"[22] and argues that one cannot reduce all of ethics to an understanding of altruism, reminds her readers that she is not "by any means opposed to every aspect of sociobiology, but only to some of its excesses."[23] Wilson and others who feel that the study of sociobiology can contribute to our understanding of the origin of culture and some of the possible tendencies to which our genes predispose us do not make the same mistake as many of their nineteenth-century predecessors, that is, in confusing the discussion of the origin of morality with its justification. If anything, they stress an opposite point: that an understanding of our genetic endowment may show us problems we need to confront. If a gender-based tendency toward aggression in humans exists, and if the roots of other behaviors are deeper than the environmental influences that are said to produce them, then stronger measures than environmental modification may be needed to control them. This tilt toward "nature" in the nature/nurture debate has been a source of much acrimonious contention and has often warped the discussion. Attacks on sociobiology, for example, have stressed alleged racist and sexist readings, and critics of sociobiology have concentrated on what they believe are its inherently conservative biases. But just as Social Darwinism in the nineteenth century took a liberal as well as a conservative form, so, too, the "data" of sociobiology can be used to justify radically different social programs.[24] These ideological skirmishes, unfortunately, often have diverted attention from other more serious conceptual issues. For if the origin of ethical codes is not what justifies them, what does? If our genes contain directions that were useful to ancient man and are maladaptive now (especially in liberal environments like Cambridge, Mass.), are we justified in labeling them "bad"? How exactly do we determine which natural dispositions are "good" and which "bad"? These issues have been principal stumbling blocks for earlier versions of evolutionary ethics.

The appearance of numerous books with sensational popularized

22. Mary Midgley, "Rival Fatalisms: The Hollowness of the Sociobiology Debate," in Montagu, Sociobiology Examined, 15–38.

23. Midgley, "Gene-Juggling," 132. Midgley argued for the importance of animal behavior for philosophers and specifically mentions Wilson's work as useful to consider in Beast and Man: The Roots of Human Nature (Ithaca: Cornell University Press, 1978).

24. Degler makes this point in his recent book, In Search of Human Nature.

versions of possible evolutionary moral lessons has also clouded the discussion of the ethical value of sociobiology. But they have little bearing on the issue. Any new and exciting perspective will have its opportunists, and in our contemporary atmosphere, vulgarization has been a profitable and well-rewarded activity.[25]

Sociobiology's relevance for ethics is analogous to that of the psychoanalytic movement or of cultural anthropology. It provides a description of influences on individuals, pressures exerted by groups on moral development, and unconscious motivation, all of which need to be understood in assessing individual action. Insight into the historical development of moral codes, and even of the beliefs of historical figures, may result from investigations in the social and natural sciences. But to go beyond description, to enter the arena of the normative, that is, to say what ought to be, involves an important shift that requires justification.

Here, sociobiology provides no new basis, no new foundation, no new hope. There is, nonetheless, a wide range of opinion on the ethical and cultural significance of sociobiology. Dawkins, on the one hand, states that culture can be independent of our genes in the sense that once culture comes into existence it may evolve "in the way that it has, simply because it is advantageous to itself."[26] He believes that man has the conscious foresight and the means so that "we, alone on earth, can rebel against the tyranny of the selfish replicators."[27]

Richard Alexander, who, like Dawkins, studies animal behavior, takes a middle position and argues that "evolutionary analysis can tell us much about our history and existing systems of laws and norms, and also about how to achieve any goals deemed desirable; but that it has essentially nothing to say about what goals are desirable, or the directions in which laws and norms should be modified in the future."[28] Alexander says that evolution has a lot to say about why people do what they do (i.e., they often act to

25. Philip Kitcher distinguishes popularized from serious sociobiology in his critique, Vaulting Ambition: Sociobiology and the Quest for Human Nature (Cambridge: MIT Press, 1985).

26. Dawkins, The Selfish Gene, 214.

27. Ibid., 215.

28. Richard D. Alexander, Darwinism and Human Affairs (Seattle: University of Washington Press, 1979): 220. This book was the result of the Jessie and John Danz Lectures at the University of Washington, 1977. Alexander elaborates on his position in The Biology of Moral Systems (New York: Aldine de Gruyter, 1987).

maximize their inclusive fitness), but he is emphatic that when it comes to what people ought to do, it says "nothing whatsoever."[29] Although Alexander is inclined to believe that the existing moral systems of thought do promote individual reproductive success, he states that such a result has come about unconsciously. Moreover, and more significant, he holds that there is no warrant for using inclusive fitness maximizing behavior as a consideration in normative ethics.[30]

Wilson, on the other hand, takes a harder line that recalls the earlier traditions of Waddington and Clifford. In his first elaboration of the moral implications of evolution, On Human Nature, he suggests that even though the genetic component of human behavior is partial, "ethical philosophy must not be left in the hands of the merely wise. . . . Only hard-won empirical knowledge of our biological nature will allow us to make optimum choices among the competing criteria of progress."[31] It is a curious argument. Wilson wants to maintain a modern Darwinian picture that envisions man as structured by his genetic heritage, which determines the epigenetic rules through which he must operate. But he also claims that man can go beyond the legacy of his nature, which is "a hodgepodge of special genetic adaptations to an environment largely vanished, the world of the Ice Age hunter-gatherer."[32] Moreover, in attempting to meet the challenge of the future, "we are forced to choose among the elements of human nature by reference to value systems which these same elements created in an evolutionary age now long vanished."[33] But there is a way out. "Fortunately, this circularity of the human predicament is not so tight that it cannot be broken through an exercise of will."[34] Through a deeper knowledge of our biology we can fashion a biology of ethics, "which will make possible the selection of a more deeply understood and enduring code of moral values."[35] Wilson is claiming that a better understanding of evolution can lead us to be more in harmony with nature; thus we might harness our reli-

29. Ibid., 271.
30. Ibid., 278. For a similar point of view, also see Francisco Ayala, "The Biological Roots of Morality," Biology and Philosophy 2 (1987): 235–252.
31. Wilson, On Human Nature, 7.
32. Ibid., 196.
33. Ibid.
34. Ibid.
35. Ibid.

gious impulses and channel the sources of our most intense emotions into behaviors that are informed by the knowledge of their likely consequences and evolutionary significance. This is not genetic determinism. Wilson recognizes man's free will. He and Charles Lumsden "suggest that moral reasoning is based on the epigenetic rules that channel the development of the mind. Such reasoning appears to be ultimately dependent on the genes as well as on culture and self-conscious decision. But the rules only bias development; they do not determine ethical precepts or the necessary decisions in a fixed manner. They still require that a choice be made, and in this sense they preserve free will."[36] Each of us, no matter how well informed, must still choose.[37] However, in spite of the liberal gloss, Wilson and Lumsden emphasize the value of survival as primary. In Wilson's earlier On Human Nature, the position was more bluntly stated. "Human behavior—like the deepest capacities for emotional response which drive and guide it—is the circuitous technique by which genetic material has been and will be kept intact. Morality has no other demonstrable ultimate function."[38]

Has Wilson produced a prolegomenon for an evolutionary ethics? He does not elaborate on the specific form ethics should take, and he explicitly rejects a calculus of genetic fitness. Furthermore, he suggests that an adequate ethics will involve extensive secondary values. Indeed, a satisfactory ethics is seen as a distant goal that can be reached only via extensive research on neurophysiology, genetics, population biology, and behavioral studies. On Human Nature can be read as a first approximation, for in it he argues for certain values that he believes promote survival. However, like the "derivations" of values from Darwinian evolution in the nineteenth century, the exercise shows merely that his values are consistent with evolution. Such a vision, however, is hardly compelling. Although not as simpleminded as his critics contend, Wilson does not tell us much more than that biology should inform our choice of values and that the survival of our genes is a cardinal value. But his hope, resonable as it sounds, has little to recommend it in the twentieth century, in that it fails to recognize that genuine conflicts exist between individuals and between groups. It

36. Lumsden and Wilson, Promethean Fire, 179.
37. Ibid., 183.
38. Wilson, On Human Nature, 167.

equally neglects the way in which individuals partake in different groups, often ones that are in opposition. History is the record of different hierarchies of values, of conflicts among and within groups. To choose one set of values over another is to commit oneself to a vision of human society, a concept of what justice is, and a vision of reality that goes far beyond mere survival.

The difference, then, between what we might call the strong program and the weak program of sociobiology-informed evolutionary ethics turns on the extent to which our biological knowledge should influence our values. The weaker program, typified by Dawkins and, to a lesser extent, Alexander, argues that sociobiology will help us understand the evolution of morality and why it is a part of human society but that it provides little guidance for moral choice. The stronger program, typified by Wilson and Lumsden, argues for a Darwinian worldview, which holds not only that evolution accounts for the origins of our moral sense but also that a deeper knowledge of the specifics of our biological nature will permit us to guide the evolutionary process rationally. Since human culture has evolved more quickly than human genes, only by rational self-control will we be able to channel the natural process of human development along beneficial lines.

The strong program certainly recalls earlier formulations of the relevance of evolution for ethics; the historian, in fact, has a most difficult time in not throwing up his floppy disks and wondering whether anyone has been reading the literature on the subject from the last hundred years. Is ignorance condemning us to repeat the past? Wilson's early call to biologize the subject is a call to ignore some important lessons.

But the weaker program is a different matter. Traditionally our picture of human nature has influenced the formulation of morality. And that picture has usually been informed by the science of the day. Psychology, in particular, has been closely linked with philosophy. Although some philosophers have rejected the value of psychology for an understanding of ethics, many in the Anglo-American community have used it as data in constructing their point of view. Psychology may not provide a justification for specific ethical positions, but it suggests some of the raw material that moral philosophers need to consider. Hume, the alleged author of the is/ought distinction, for example, made considerable use of his conception of the mind in formulating an ethical position. On a more complex level, Donald Campbell, in his 1975 presidential address to the American Psychological Association, suggested that human social evolution had developed in opposition to certain biological tendencies and had to be understood in order to avoid misguided moralizing by social scientists.[39] The current state of sociobiology may be no better than a set of nested black boxes, but, along with psychoanalysis, linguistics, and anthropology, it is, at a minimum, a possible source of relevant information. In the whirlwind of rhetoric that has surrounded the discussions of sociobiology, this point has often been lost. Social scientists, instead, have reacted with typical professional fury at the suggestion that the relationship of sociobiology to their disciplines is similar to the relationship between chemistry and alchemy, or between astronomy and astrology. Also overlooked is that the essential proposal is a research program rather than a body of interpreted data. It is not that Wilson or anyone else has cracked the genome, related specific genes, groups of genes, developmental histories, or whatever, to specific human behavior, mediated by environmental influences, interactions with other humans, or the luck of the draw. The relevant information that they claim will be of interest to ethics is yet to be had. While this may be of substantial interest to funding agencies or for faculty hiring decisions in the natural sciences, it is not surprising that philosophers have tended to ignore the uproar of their obstreperous neighbors. Although Wilson's On Human Nature received a Pulitzer Prize in 1979, his claims for the ethical value of sociobiology are ignored by most of the philosophical community. There are some notable exceptions. Mary Midgley, who was initially supportive of the potential value of animal behavior for philosophy, has come out against what she sees as a new Darwinian myth that threatens to engulf rational inquiry.[40] She fears that some writers are "obsessed by a picture so colourful and striking that it numbs thought about the evidence required to support it."[41] And her point is well taken, especially in light of the popular sociobiology that has made such speculative claims. An established picture of "human nature" from which to derive useful lessons is far away.

39. See Donald T. Campbell, "On the Conflicts Between Biological and Social Evolution and Between Psychology and Moral Tradition," American Psychologist 30 (1975): 1103–1126.

40. Mary Midgley, Evolution as a Religion: Strange Hopes and Stranger Fears (London: Methuen, 1985).

41. Ibid., 5.

In contrast, a few philosophers, like Michael Ruse and the late John Mackie, have been attracted to the potential of sociobiology to provide new approaches in ethics. They are acutely aware of the many problems that invalidate earlier attempts. Ruse, in fact, critically reviews the history of evolutionary ethics and dismisses the major arguments from Spencer to Wilson. But not completely, for although he rejects most of Wilson's program for an evolutionary ethics, he believes that Wilson has raised at least one major issue that may yet prove to be of substantial value: "that moral claims must be explained by factual evolutionary claims."[42] Ruse makes this point, tellingly, in a book that adumbrates a neo-Darwinian worldview, Taking Darwin Seriously: A Naturalistic Approach to Philosophy (1986). According to Ruse, there is no objective foundation on which to construct ethics. He finds uncompelling contemporary approaches, such as emotivism. Instead he pursues his Darwinian vision, and it is to sociobiology that he turns to demonstrate the origin of altruism by way of individual natural selection. Ruse argues that morality needs to be considered among "the genetically based dispositions to approve of certain courses of action and to disapprove of other courses of action."[43] These dispositions are biological adaptations. He provocatively states that perhaps "morality is a collective illusion foisted upon us by our genes."[44] By this he means that ethics is a subjective enterprise and that Darwinism is the best approach we have for understanding why we embrace the moral codes we do. His view is a type of moral subjectivism, which stakes its hope on evolution to provide some guidance, or at least some emotional reinforcement of the values we hold. Similarly, Mackie has argued that no objective basis for morality exists and that biological concepts like "evolutionary stable strategies" may be helpful in discussing practical morality.[45] Ruse and Mackie stress that our knowledge of biological evolution provides information for informed decisions and that

42. Ruse, Taking Darwin Seriously, 100.
43. Ibid., 221.
44. Ibid., 253.
45. See J. L. Mackie, Ethics: Inventing Right and Wrong (London: Penguin Books, 1977), and his articles, "The Law of the Jungle: Moral Alternatives and Principles of Evolution," Philosophy 53, no. 206 (1978): 455–464, "Genes and Egoism," Philosophy 56, no. 218 (1981): 553–555, and "Co-operation, Competition, and Moral Philosophy," in J. L. Mackie, Persons and Values: Selected Papers. Vol. 11, ed. John Mackie and Penelope Mackie (Oxford: Oxford University Press, 1985): 152–169.

we have moral sentiments that incline us in these directions. They fail, however, to tell why we are obliged to follow those sentiments. A moral choice, after all, goes beyond merely following what we feel is correct.

A few writers take a stronger position and argue that an objective basis exists for justifying evolutionary ethics. They see it, like the quest for the firebird in Russian mythology, as a possible solution to the elusive problem of an acceptable ethics in the twentieth century. Carla Kary, for example, argues that if Wilson is correct, we have a testable view of human nature that can be used as a ground for ethics. Her point is that "ethicists can offer satisfying, normative recommendations by showing how one ethical theory is a better fit to the genetic constraints of Human Nature than others, and hence why that ethical theory will be the most fruitful theory to adopt."[46] Kary wants to ground ethics in terms of our potential knowledge of genetic constraints. But demonstrating that an ethical position is consistent with known biological facts and theories is vastly different from showing that an ethical position follows from them. If sociobiology does indeed give us information about human nature then one would expect philosophers to utilize that knowledge in any system that depends on a vision of human nature. But views such as Kary's rest on a highly simplified notion of ethics. Moral philosophy, as Midgley states, traditionally attempts "to understand, clarify, relate, and harmonize so far as possible the claims arising from different sides of our nature."[47]

The ideas of Robert Richards, an intellectual historian who has made a serious foray into these treacherous waters, are similar to Kary's. Richards argues for the theoretical possibility of a valid evolutionary ethics if we can show empirically that man has evolved a moral sense that "in appropriate circumstances move[s] the individual to act in specific ways for the good of the community."[48] His position rests on showing that if we can demonstrate altruism as part of man's nature, then we are in a position to construct an argument along the lines of those who argue from man's nature to normative injunctions. Richards uses a hypothetical sce-

46. Carla Kary, "Sociobiology and the Redemption of Normative Ethics," The Monist 67, no. 2 (1984): 163.
47. Midgley, Beast and Man, 169.
48. Richards, Darwin and the Emergence of Evolutionary Theories, 603.

nario to make his point. He assumes that early man evolved in small groups and that "they often acted to benefit other community members without expectation of reciprocation and that they prized such behavior in others."[49] The origin of such behavior could have resulted from kin selection and natural selection on small groups. In other words, early kin selection would give rise to individuals who acted altruistically not only for the benefit of close relatives but for the benefit of the group. Like Darwin, Richards then imagines group selection favoring such communities and believes that the extension of altruistic behavior to community members avoids the charge that individuals are acting out of (genetic) self-interest. Moreover, Richards contends that such a scenario provides an argument for evolutionary ethics without committing any logical error.

But even such a circumscribed position is highly problematic. For one thing, it relies on the empirical claim that the action of group selection in human history favored communities fostering individuals who acted altruistically. Group selection, however, is accepted by few scientists today. More important, Richards treats ethics as a subject that is encompassed by the notion of altruism. That is surely a simplification and overlooks the enormous moral conflicts to which any ethical system is expected to be relevant. Of equal seriousness, his sketch sidesteps the issue of justifying the foundation of ethics as "community good." Richards defines "being moral" as acting for the community good and then states that it may be possible to show empirically that evolution has resulted in animals that have just such characters.[50] But, as Sidgwick would undoubtedly have asked, why should we accept community good as the highest good? Richards tries to head off such criticism by claiming that the justification of moral principles "must ultimately lead to an appeal to the beliefs and practices of men, which of course is an empirical appeal. So moral principles ultimately can be justified only by facts."[51] Philosophers since Sidgwick, however, have contended that ethics should provide the grounds for accepting common moral principles. Although there have been a number

49. Robert Richards, "Dutch Objections to Evolutionary Ethics," Biology and Philosophy 4 (1989): 331.
50. Ibid., 623–624.
51. Richards, Darwin and the Emergence of Evolutionary Theories, 619.

of articles inspired by Richards's argument in Biology and Philosophy, not many philosophers have found his hypothetical evolutionary ethics persuasive.[52]

More guarded than Richards or Kary are Neil Tennant and Florian von Schilcher, who present a careful and balanced discussion of both the biology and the philosophy of sociobiology. They conclude that sociobiology "does not offer any criteria of value or normative laws,"[53] but they regard the weaker program, nonetheless, as having considerable worth.

> Even if our theories today indicate no definite policy directions, at least we should be beginning to appreciate the importance of a particular kind of consideration. This is that our basic behavioral biology now finds itself in newly created niches in which it may be ill adapted. The hope remains that we shall be able to chart the pitfalls and avalanches on our epigenetic landscape that threaten those acting in accordance with the wrong universalized maxims. This, indeed, is what determines those maxims as wrong. When Kant enjoins us to ask whether the maxim by which we are acting could consistently be willed to become a universal maxim, the sorts of possibilities considered when judging thus of consistency must be the physical possibilities admitted by the laws and facts of nature. Among these are the laws of genetics, and the fact of evolution past and future. If policymakers were to become more aware of these laws and facts, then evolutionary theory would finally become a source of grounded wisdom about our own nature.[54]

Will an evolutionary perspective on human behavior turn out to have value for ethics? By way of establishing a foundation or

---

52. See Robert Richards, "A Defense of Evolutionary Ethics," Biology and Philosophy 1, no. 3 (1986): 265–293. Responses by Camilo Cela-Conde, Alan Gewirth, William Hughes, Laurence Thomas, and Roger Trigg are in the same issue, as is Richards's "Justification Through Biological Faith: A Rejoinder." See also Richards, "Dutch Objections to Evolutionary Ethics"; Bart Voorzanger, "No Norms and No Nature—The Moral Relevance of Evolutionary Biology," Biology and Philosophy 3 (1987): 253–270; Stephen Ball, "Evolution, Explanation, and the Fact/Value Distinction," Biology and Philosophy 3 (1988): 317–348; Patricia Williams, "Evolved Ethics Re-Examined: The Theory of Robert J. Richards," Biology and Philosophy 5 (1990): 451–457; William Rottschaefer, "Evolutionary Naturalistic Justification of Morality: A Matter of Faith and Works," Biology and Philosophy 6 (1991): 341–349; Michael Bradie, "Darwin's Legacy," Biology and Philosophy 7 (1992): 111–126.

53. Florian von Schilcher and Neil Tennant, Philosophy, Evolution and Human Nature (London: Routledge and Kegan Paul, 1984): 165. Also see Neil Tennant, "Evolutionary versus Evolved Ethics," Philosophy 58, no. 225 (1983): 289–302.

54. von Schilcher and Tennant, Philosophy, Evolution and Human Nature, 163. A recent study that extends the discussion of the potential value of an evolu-

justification, recent constructions have not improved on earlier flawed attempts to derive an ethics from evolution. But perhaps if philosophers develop an ethical theory or ethical dialogue that is nonfoundationalist, evolutionary considerations may enter the philosophical arena. To an extent, both Ruse and Richards suggest this approach. After all, the search for a foundation for ethics has hit on hard times. Since Sidgwick confessed to his lack of success in systematizing morality without reference to premises taken on faith, Anglo-American philosophy has made little progress. Sixty years ago, C. D. Broad warned about the unlikelihood of constructing a unified rational ethic.

> One lesson at least has been taught us so forcibly by our historical and critical studies in the theory of Ethics that we ought never to forget it in future. This is the extreme complexity of the whole subject of human desire, emotion, and action; and the paradoxical position of man, half animal and half angel, completely at home in none of the mansions of his Father's house, too refined to be comfortable in the stables and too coarse to be at ease in the drawing-room. So long as we bear this lesson in mind we can contemplate with a smile or a sigh the waxing and waning of each cheap and easy solution which is propounded for our admiration as the last word of "science." We know beforehand that it will be inadequate; and that it will try to disguise its inadequacy by ignoring some of the facts, by distorting others, and by that curious inability to distinguish between ingenious fancies and demonstrated truths which seems to be the besetting weakness of men of purely scientific training when he steps outside his laboratory. And we can amuse ourselves, if our tastes lie in that direction, by noticing which well-worn fallacy or old familiar inadequacy is characteristic of the latest gospel, and whether it is well or ill disguised in its new dress.[55]

Much twentieth-century ethics has avoided the issue of a unified rational ethics by focusing attention on the meaning of ethical terms, or by elaborating theories of ethics that take as their starting point the position that ethical statements are not knowledge claims but rather emotive or prescriptive statements. Although a few philosophers like Alasdair MacIntyre suggest that the attempt

---

tionary perspective on human nature is James Q. Wilson, *The Moral Sense* (New York: Free Press, 1993).

55. C. D. Broad, *Five Types of Ethical Theory* (London: Kegan Paul, Trench, Trubner, 1930): 284.

to construct a rational ethics is doomed unless we make a commitment to a general worldview and once again take metaphysics seriously, a general malaise seems to hover over the entire field. For others, like Bernard Williams, for example, ethics seems to be overtaxed by unrealistic demands. Along with Stuart Hampshire, he doubts whether an adequate rational system of ethics can be successfully constructed. Pragmatists like Richard Rorty have written of returning to Dewey for guidance, but more recently, they have drawn on Nietzsche and Derrida to sketch an ethical position that circumvents any unified theory. Others, like G. J. Warnock, are more optimistic about the possibility of constructing a rational ethics, which even if it does not resolve all dilemmas, at least can provide general guidance for the amelioration of the human situation that he contends "is inherently such that things are liable to go badly."[56] Similarly, John Rawls's widely read book, Theory of Justice (1971), proposes an interesting contractual account of ethics that seeks to establish principles to guide social and political conduct.[57] What is significant for our story about the current literature is that evolutionary ethics is rarely mentioned, even by those who are optimistic about the future of research in ethics.[58]

Whether it is the death throes of ethical philosophy that we are witnessing or merely a quiet phase of its exciting metamorphosis is not the issue here. Rather, it is that for most professional philosophers, even those with a preference for a naturalistic ethics, sociobiology has not increased the plausibility of contemporary evolutionary ethics over its earlier versions. Philosophers are skeptical of any insights ostensibly gained from genetics and feel that the burden of proof rests with the biologists. And just as the evolutionary ethics of the 1940s lacked a fully formulated exposition and therefore remained unconvincing, so, too, does the newest pro-

---

56. G. J. Warnock, The Object of Morality (London: Methuen, 1971): 17. Also see Bernard Williams, Ethics and the Limits of Philosophy (Cambridge: Harvard University Press, 1985); Stuart Hampshire, Morality and Conflict (Oxford: Basil Blackwell, 1983); and Richard Rorty, Contingency, Irony, and Solidarity (Cambridge: Cambridge University Press, 1989). All pragmatists have not gone the route of Rorty. See, for example, Paul Kurtz, Philosophical Essays in Pragmatic Naturalism (Buffalo: Prometheus Books, 1990).

57. John Rawls, A Theory of Justice (Cambridge: Harvard University Press, 1971).

58. See, for example, the overview of current research in ethics, Stephen Darwell, Allan Gibbard, and Peter Railton, "Toward Fin de siècle Ethics: Some Trends," Philosophical Review 101, no. 1 (1992): 115–189.

gram for an evolutionary ethics look not only "conjectural" but also unpromising as a theory of ethics.[59] A reflection of the lack of the impact of evolutionary ethics can be seen in the relative indifference of the antievolutionist movement. Barely a reference to evolutionary ethics is to be found in the sizable literature that fundamentalists have published in this century. Whether the scientific study of human behavior will contribute to our understanding of the nature of man and whether such understanding will augment our ethical theories are open questions.

---

59. Kenneth Bock writes, "The failure of evolutionism lies in the fact that it was just what its earlier exponents called it—'conjectural history.'" See Bock, "Theories of Progress and Evolutionism," in Werner Cahnman and Alvin Boskoff, eds., Sociology and History: Theory and Research (New York: Free Press, 1964): 36.

9. Recapitulation, Lessons, and Queries

Does biology hold the key to human nature? Even a casual reading of the current literature on the subject suggests that some scientists, humanists, and social scientists think so. From articles in highbrow periodicals like the Times Literary Supplement to those in mass market magazines like Newsweek, the message is clear: biology is poised to contribute significantly to our understanding because of discoveries in research on neurobiology, the human brain, human behavior, and the human genome. There are voices that dissent. Many humanists and social scientists object to what they see as a facile reduction of their subject to simplistic explanations. Even more worrisome to critics is the concern that far from revealing secrets of human nature, scientific investigators are merely reading their social attitudes into nature. These critics have no difficulty finding flagrant abuses in the past, such as the history of the eugenics movement, to remind us of how easy it is to justify social opinion by projecting our values onto nature.

Historians like Daniel Kevles and Evelyn Fox Keller have contributed to our understanding of how easily human values can infiltrate our picture of nature. This study has examined several writers who have attempted to construct an evolutionary ethics, or at least to inform ethics with evolutionary history. As we have seen, their story has not been simply a reading into evolution of human values but has involved many different interpretations spread out over three distinct periods. Among the first to signal the significance of evolution for ethics was Charles Darwin. After having pondered the origin of man's moral sentiment for many years, he published his speculations in the Descent of Man. Darwin thought that by approaching the topic from the perspective of natural history, new light would be shed on this perennial problem. Ethnology had suggested to him that a set of universal ethical norms existed among the diverse peoples of the world and that, in addition, human moral development exhibited a range of variation from simple to complex. Since Darwin envisioned human evolution on a continuum with animal evolution, quite naturally, he saw the moral sense as further development of the social instincts. This naturalistic account engendered enthusiasm in people like Leslie Stephen and William Kingdon Clifford, who saw it as a new scientific foundation for ethics compatible with the utilitarian spirit and high moral tone of the period. Stephen, ardently seeking a replacement for religious ethics, saw as his greatest accomplishment the construction of an evolutionary ethics. Like Darwin and Clifford, Stephen did not seek specific moral guidance from evolution. Like most of his contemporaries, he believed he knew right from wrong and wanted to mold society in conformity with his values. But the old religious sanctions that had provided a justification for his beliefs no longer satisfied. He hoped the new fruitful scientific approach might serve to fill this void.

Similarly, Herbert Spencer looked to the past for justifying his visions of the future. Spencer concerned himself with how men ought to live, and he believed that the entire course of history—cosmic, biological, and social—pointed to evolutionary principles that could provide a guide. He distinguished, however, "Absolute" from "Relative" ethics, that is, those rules that applied to the ideal state and those that existed on the long, hard road to it. Catching the spirit of the age, he embedded his ethical viewpoint in a vast philosophical overview and attracted a significant audience. Unlike Darwin, who avoided publishing on moral issues, Spencer eagerly gave advice to his generation, stressing the inviolability of the individual and the benefits of industrial society. And he was appreciated by a broad range of readers. Secularists like Clifford used Spencer to justify their ideas, as did religious reformers like John Fiske and Henry Drummond.

But the enthusiasm for an evolutionary interpretation of ethics was not universal. Even among the inner circle of evolutionists critics existed. Thomas Henry Huxley and Alfred Russel Wallace strongly objected to the extension of evolution to the realm of morality. Huxley, especially, popularized reasons to give pause to those who were attracted to reading moral lessons from evolution. His critique, stressing the opposition of cultural to biological evolution, was more devastating than Wallace's, whose naive, spiritualist sympathies tended to discredit his objections. But Huxley had no alternative to evolutionary ethics. His agnosticism, although brave in the face of the dark clouds that threatened the last decade of the nineteenth century, provided little psychic comfort for general readers.

Of more importance than the defection of two leaders from the Darwin camp were philosophers who examined evolutionary ethics for its adequacy as a systematic, rational account and found it seriously wanting. Insight into the origin of morality, although of great interest to Darwin, held little weight with philosophers who sought its justification. Henry Sidgwick, for example, pursued the most developed system of evolutionary ethics, that of Herbert Spencer's, with a relentless fury, tempered only by his Cambridge civility.

In spite of the withering reception given evolutionary ethics by the philosophical community, popular writers continued to elaborate on evolutionary themes in their discussions of moral issues and ethics. Many of these moved far beyond anything Darwin or Spencer would have endorsed, such as Benjamin Kidd's vision of a spiritual evolution led by women. Interesting and original as some of these "evolutionary" scenarios were, none of them were able to attract extended intellectual debate or acceptance. Aside from its pedagogic value in demonstrating fallacious ethical arguments, philosophers on both sides of the Atlantic generally had little use for evolutionary ethics.

The theory of evolution, both in its narrower Darwinian mode and in its more expansive Spencerian one, initially served as an inspiration for writers to develop a new perspective to fill an ethical void caused by the loss of faith in religious sanction as a foundation for ethics. The second episode in the history of evolutionary ethics followed from the formulation of a reborn and more powerful Darwinian interpretation of evolution, the modern synthesis, and from scientific innovations in psychology that suggested a new approach to understanding man's sense of obligation. Although the need for a new foundation for ethics had been questioned by professional philosophers like G. E. Moore and A. J. Ayer who believed the problem was solved, or was a pseudo-issue, the general public considered the issue significant. Whether evolutionary models could supply an adequate solution, however, was called into question by a shift away from evolutionary explanations in many disciplines. Anthropology, psychology, and even subdisciplines in biology sought new fruitful perspectives to solve their problems. But interest in Freud's theory of the subconscious powers of the mind and renewed excitement over the power of Darwin's original selectionist emphasis in biology gave hope to some that evolution would yet provide a base on which to build a moral vision. Or, at least one on which to ground a moral vision they held. Writers like Julian Huxley, in the shadow of first one and then another devastating continental war with global repercussions, sought in the evolutionary process a secular faith to justify their hopes for humanity. Others like George Gaylord Simpson held that the evolutionary forces that had created man did not imply any particular moral code but that human ethical standards ought to be relative to man's evolution.

Philosophers could have dusted off nineteenth-century critiques on evolutionary ethics in response, but most did not think the musings of scientists important enough to make the effort. Emeritus professors, advocates of scientism, and well-meaning but philosophically naive colleagues were not sufficiently challenging targets for a professional rebuttal. And that is probably where the issue would have remained had it not been for the stunning boost given evolutionary explanations of human behavior by the synthesis of genetics, behavior, and evolution in the 1970s. Wilson's Sociobiology was central, as well as symbolic of this burst of activity, and his writings have been crucial in the discussion.

The field of sociobiology inspired a new, third episode in the history of evolutionary ethics. Some lessons were learned from earlier, unsuccessful attempts. The new arguments are well informed by the latest biology, and the teleology that had marked the outmoded Spencerian philosophy has all but disappeared. We are in a truly Darwinian landscape. Moreover, the patent circularity of Huxley or Waddington is now rarely encountered in discussions of contemporary evolutionary ethics. Instead, more sophisticated discussions of inherited "tendencies," biological aspects of human nature, or lessons to be learned from our evolutionary past are advanced. But the results of the new evolutionary ethics have not been encouraging. The fundamental problems dogging earlier attempts remain unresolved: the oversimplification of the conception of ethics; the lack of an independent justification for values; the lack of a rational justification of one's obligation to comply with those values; and the enormous gulf between actions that promote survival and actions that are deemed moral. Although an understanding of the evolutionary significance of morality and knowledge of the behavioral tendencies of humans may some day provide relevant data for ethics, to date our understanding has not provided an adequate base on which to build a system of ethics. Can we glean anything positive from this story, other than to "just say no" to evolutionary ethics? I think that there are several lessons that are instructive. First, this history illustrates the powerful invitation that the theory of evolution has extended toward the development of an absolute, objective basis for ethics. Evolution is a theory that scientists have developed far beyond its original scope and that now unifies the biological sciences. It makes sense of animal form, distribution, and the fossil record, as well as provides a perspective from which to view behavior, physiology, and ecology. If the social sciences are what Comte and others since him have believed them to be, that is, the scientific study of man, then is not the extension of evolutionary theory to mankind natural? Perhaps. It depends on what questions are asked and how they are framed. Evolutionary biology has been highly successful in accounting for the phenomena of the living world, but it has had its failures and dead ends. The effort to construct speculative phylogenies, for example, bedeviled morphology for years after Darwin, and it took nearly a revolution in the biological sciences at the turn of the century for a more profitable research program to take hold. The study of human evolution has been illuminating and exciting in this century, and hardly a year goes by now that we do not have to revise our picture in light of new evidence. But just because evolution tells us of man's origins, must we believe also that it opens up the hope of illuminating our prospects? Traditionally, accounts that reveal origins chart destiny. The invitation to move from the physical evolution of man to an evolutionary account of social behavior and its norms is a tempting challenge.

Second, the history of evolutionary ethics shows that the theory of evolution, although powerful in helping us to understand our biological background, and clearly inviting extension into the human domain, has not really helped us much with the formulation or justification of human values. In a way, arguments over the significance of evolution for ethics are reminiscent of discussions by logical empiricists in the 1960s on the reduction of biological theories. An objection to that goal, as then defined, was that even if we could "reduce" biological phenomena by way of extensive correspondence laws to physics, of what use would the exercise be? Many people felt not much, in that it would not be highly informative or predictive; that is, it would not tell us much that we did not already know. Would we foretell the existence of a creature such as the duck-billed platypus from the laws of quantum mechanics?[1] In a similar way, it is unlikely that we could shed light on a moral dilemma—for example, should I have an abortion even though I am a Catholic?—by reference to inclusive fitness, or a broad, evolutionary understanding of the origins of morality.

A third lesson from our story concerns why evolutionary ethics has not worked. In part, any answer to that question must consider what is expected of a theory of ethics. There is, for instance, an old distinction between morality and ethics that must be considered. Morality is a body of specific rules or guides to behavior, whereas ethics is the justification and systematic arrangement of those rules. Philosophers have repeatedly pointed out that as a guide, evolutionary considerations are too broad to be of value. Similarly, as a foundation, they either beg the question or are so general as to be meaningless. Leslie Stephen's attempt to anchor ethics in their evolutionary origins, Julian Huxley's use of Freud to support his lofty vision of evolution having resulted in the power and responsibility of man directing destiny, Edward O. Wilson's contention that biology holds the key to an ethical reexamination of human conduct and attitudes—all these efforts reduce to a reading in of values to the cosmic process. Survival and/or adaptation are themselves ethically neutral. Merely increasing copies of our genes has little moral appeal as the literature of dystopias and science fiction, replete with accounts of alternative societies that promote gene replication at the "loss of humanity," indicate.

That evolutionary ethics has had such a long run is certainly interesting given its problematic nature. Why has it been revived so often in the face of such daunting criticism? Part of understanding this story requires a recognition of the unresolved crisis in Anglo-American ethical thought dating from the 1870s. The loss of an

---

1. I do not mean to imply that the issue of reductionism was so clear-cut. The existence of a continuing debate over reduction certainly demonstrates the complexity of the subject. While the original logical empiricist program for theory reduction, as exemplified by Ernest Nagel, may be dead, the broader issue of theory reduction and replacement is alive and generating Ph.D.'s.

agreed upon ethical foundation has been a source of concern and a central focus for many thinkers. It has been a multidimensional problem. Some philosophers have even suggested that further efforts are in vain; a rational foundation for ethics in the modern world is an impossibility. But if so, they have not provided an alternative outlet for discussing perennial moral issues.[2] And until they (or some other group) do, it is reasonable that other paths are explored, profitably or not. For the topic of ethics, although formally a part of philosophy, is of interest to a much wider audience. The desire for a foundation for ethics has been motivated by different emotional and intellectual needs. Members of the scientific community have been drawn to the subject by their desire for an account of the origin of values and an analysis of their truth value. Scientism, the position that natural science is the "most valuable part of human learning," flourishes in the intellectual void left by philosophers and encourages a "scientific" approach to the subject.[3] Not surprisingly, numerous biologists have regarded the theory of evolution as a potential source of enlightenment on the subject.

This has been especially so among those scientists who have had strong interests in social reform. Figures like Julian Huxley saw in the theory of evolution the potential for a new mythical origin of values. The criticism that evolutionary ethics was so general as to be amenable to an extremely wide range of moral positions posed less of a problem for someone like Huxley who knew what he wanted to justify than to someone like Sidgwick who wanted to construct a rigorous and systematic rational justification for accepted morality. This is not to imply that all advocates of evolutionary ethics have had social agendas, certainly not that they shared a single one. But the widespread attraction that evolutionary ethics has held for many authors with interest in social change

2. MacIntyre captures the malaise in moral philosophy when he wrote, "Very little has been written that philosophers working outside moral philosophy are likely to recognize as characterized by philosophical inventiveness of a kind that could be important to their own inquiries." See Alasdair MacIntyre, "Moral Philosophy: What Next?" in Stanley Hauerwas and Alasdair MacIntyre, eds., Revisions: Changing Perspectives in Moral Philosophy (Notre Dame: University of Notre Dame Press, 1983): 4.

3. An interesting introduction to scientism in philosophy is Tom Sorell, Scientism: Philosophy and the Infatuation with Science (London: Routledge, 1991). Sorell briefly discusses Darwinian approaches to ethics and criticizes them as inadequate.

suggests that historians need to look closely at how and why scientific theories come to be extended to the social sciences.

But social reform has not been the only motivating factor in the quest for an evolutionary ethics. Some have been drawn to an attempt to unify our diverse intellectual domains under a single general concept. To what extent have such attempts (natural selection, economic determinism, God's will) been useful? The history of philosophy is littered with the fossil remains of such constructions: medieval reconciliations of pagan learning with Christianity, the great dreams of reason of the seventeenth century, architectonic systems of the Enlightenment, and German romantic rational reconstructions of reality stand as monuments and reminders of past conceptions of unified knowledge. Is the attempt to extend evolutionary insights into the realm of values an example of the quest for unified knowledge? Can we live better without such crusades? Should we put away our Leibnizian playthings and try to fashion a pluralistic worldview as we enter the twenty-first century?

A humbler attitude might help us avoid facile readings into nature of our own values. Similarly, a comparative study of the history of evolutionary ethics might instruct us on the extent to which we in the past have projected our own opinions onto nature. Perhaps a look at the Russian, German, and French intellectual traditions would demonstrate the cultural-bound nature of evolutionary ethics in a more striking fashion than does this inquiry into only the Anglo-American tradition. But that will have to await future research on the subject.

The history of evolutionary ethics is a topic that reflects both on the theory of evolution and on ethics. Because it explores man's origins, evolution extends an invitation, and given the state of ethics, that invitation has been attractive. In coming decades philosophers may judge the construction of ethical systems to be impossible. The entire field of ethics itself may come to be regarded as a pseudo-subject like astrology. But perhaps a different approach in ethics will prove to be fruitful. Historians, however, should not try to predict the course of philosophy, or they cease to be historians. At best they can point out a century-long frustration over the state of ethics and in the case of evolutionary ethics, the significantly interesting but dismal track record of this approach as a method of resolution.

Bibliography

Aeschliman, Michael. The Restitution of Man: C. S. Lewis and the Case Against Scientism. Grand Rapids: William B. Eerdmans, 1983.

Alexander, Richard D. The Biology of Moral Systems. New York: Aldine de Gruyter, 1987.

———. Darwinism and Human Affairs. Seattle: University of Washington Press, 1979.

Alexander, Samuel. "Natural Selection in Morals." International Journal of Ethics 11, no. 4 (1882): 409–439.

———. Philosophical and Literary Pieces. Edited by John Laird. London: Macmillan, 1939.

———. Space, Time and Deity. London: Macmillan, 1920.

Allen, Garland. Life Science in the Twentieth Century. New York: John Wiley and Sons, 1975.

Allen, Grant. "The Gospel According to Herbert Spencer." Pall Mall Gazette 50, no. 7832 (April 26, 1890): 1–2.

Annan, Noel. Leslie Stephen: The Godless Victorian. New York: Random House, 1984.

———. "The Intellectual Aristocracy." In J. H. Plumb, ed., Studies in Social History: A Tribute to G. M. Trevelyan. London: Longmans, Green, 1955. Pp. 241–287.

Ashton, Rosemary. G. H. Lewes: A Life. Oxford: Oxford University Press, 1991.

Atkinson, J. W. "E. G. Conklin on Evolution: The Popular Writings of an Embryologist." Journal of the History of Biology 18, no. 1 (1985): 31–50.

Austin, J. L. Philosophical Papers. Oxford: Oxford University Press, 1961.

Ayala, Francisco. "The Biological Roots of Morality." Biology and Philosophy 2 (1987): 235–252.

Ayer, A. J. Language, Truth and Logic. London: Victor Gollancz, 1936.

———. Philosophy in the Twentieth Century. New York: Random House, 1982.

Ayer, A. J., ed. The Humanist Outlook. London: Pemberton, 1968.

Badcock, C. R. The Psychoanalysis of Culture. Oxford: Basil Blackwell, 1980.

Bagehot, Walter. Physics and Politics, or Thoughts on the Application of the Principles of "Natural Selection" and "Inheritance" to Political Society. London: H. S. King, 1872.

Bain, Alexander. Autobiography. London: Longmans, Green, 1904.

———. "The Data of Ethics." Mind 4, no. 16 (1879): 561–569.

———. Mental Science: A Compendium of Psychology, and the History of Philosophy. Designed as a Text Book for High Schools and Colleges. New York: D. Appleton, 1868. (English edition entitled Mental and Moral Science. London: Longmans, Green, 1868.)

———. "Mr. Sidgwick's Method of Ethics." Mind 1, no. 2 (1876): 179–197.

Baldwin, J. Mark. "The Influence of Darwin on Theory of Knowledge and Philosophy." Psychology Review 16 (1909): 207–218.

Baldwin, Thomas. G. E. Moore. London: Routledge, 1990.

Balfour, Arthur James. The Foundations of Belief: Being Notes Introductory to the Study of Theology. 8th ed. London: Longmans, Green, 1901. (First edition, 1895.)

Ball, Stephen. "Evolution, Explanation, and the Fact/Value Distinction." Biology and Philosophy 3 (1988): 317–348.

Balmforth, Ramsden. "The Influence on Darwinian Theory on Ethics." International Journal of Ethics 21, no. 4 (1911): 448–465.

Bannister, Robert. Social Darwinism: Science and Myth in Anglo-American Thought. Philadelphia: Temple University Press, 1979.

———. Sociology and Scientism: The American Quest for Objectivity, 1880–1940. Chapel Hill: University of North Carolina Press, 1987.

Banton, Michael, ed. Darwinism and the Study of Society: A Centenary Symposium. London: Tavistock Publications, 1961.

Barratt, Alfred. Physical Ethics or the Science of Action: An Essay. London: Williams and Norgate, 1869.

Barrett, Paul H., ed. Metaphysics, Materialism, and the Evolution of Mind: Early Writing of Charles Darwin. Chicago: University of Chicago Press, 1974.

Barrett, Paul, et al. Charles Darwin's Notebooks, 1836–1844. Ithaca: Cornell University Press, 1987.

Bartholomew, Michael. "Huxley's Defence of Darwin." Annals of Science 32 (1975): 525–535.

Barton, Ruth. "'An Influential Set of Chaps': The X-Club and Royal Society Politics 1864–85." British Journal for the History of Science 23 (1990): 53–81.

———. "John Tyndall, Pantheist: A Rereading of the Belfast Address." Osiris, 2d ser., 3: 111–134.

Bendall, D. S., ed. Evolution from Molecules to Man. Cambridge: Cambridge University Press, 1983.

Benn, Alfred. "The Relation of Ethics to Evolution." International Journal of Ethics 11, no. 1 (1900): 60–70.

Bennett, J. H., ed. Natural Selection, Heredity, and Eugenics: Including Selected Correspondence of R. A. Fisher with Leonard Darwin and Others. Oxford: Oxford University Press, 1983.

Benson, Keith, Jane Maienschein, and Ronald Rainger, eds. The Expansion of American Biology. New Brunswick: Rutgers University Press, 1991.

Berman, Milton. John Fiske: The Evolution of a Popularizer. Cambridge: Harvard University Press, 1961.

Bibby, Cyril. Scientist Extraordinary: The Life and Scientific Work of Thomas Henry Huxley, 1825–1895. Oxford: Pergamon Press, 1972.

————. T. H. Huxley: Scientist, Humanist and Educator. London: Watts, 1959.

Bird, Graham. William James. London: Routledge and Kegan Paul, 1986.

Bixby, James Thompson. The Crisis in Morals: An Examination of Rational Ethics in the Light of Modern Science. Boston: Roberts Brothers, 1891.

Bledstein, Barton. The Culture of Professionalism: The Middle Class and the Development of Higher Education in America. New York: W. W. Norton, 1976.

Blythe, Ronald. The Age of Illusion: England in the Twenties and Thirties, 1919–1940. London: Hamish Hamilton, 1963.

Boakes, Robert. From Darwin to Behaviorism. Cambridge: Cambridge University Press, 1984.

Boas, Franz. The Mind of Primitive Man. Rev. ed. New York: Free Press, 1963. (First edition, Macmillan, 1911.)

Bock, Kenneth. "Theories of Progress and Evolutionism." In Werner Cahnman and Alvin Boskoff, eds., Sociology and History: Theory and Research. New York: Free Press, 1964. Pp. 21–41.

Bollen, Paul F., Jr. American Thought in Transition: The Impact of Evolutionary Naturalism, 1865–1900. Chicago: Rand McNally, 1969.

Booth, (Gen.) William. In Darkest England, and the Way Out. New York: Funk and Wagnalls, 1890.

Bowler, Peter. The Eclipse of Darwinism: Anti-Darwinian Evolution Theories in the Decades around 1900. Baltimore: Johns Hopkins University Press, 1983.

————. Evolution: The History of an Idea. Berkeley, Los Angeles, and London: University of California Press, 1984.

————. "Scientific Attitudes to Darwinism in Britain and America." In David Kohn, ed., The Darwinian Heritage. Princeton: Princeton University Press, 1985. Pp. 641–681.

Boyd, Robert, and Peter Richerson. Culture and the Evolutionary Process. Chicago: University of Chicago Press, 1985.

Bradie, Michael. "Darwin's Legacy." *Biology and Philosophy* 7 (1992): 111–126.

Bradley, F. H. Ethical Studies. Oxford: Oxford University Press, 1927. (First edition, 1876.)

Brennan, Bernard. The Ethics of William James. New York: Bookman Associates, 1961.

Brettschneider, Bertram. The Philosophy of Samuel Alexander: Idealism in "Space, Time, and Deity." New York: Humanities Press, 1964.

Breuer, Georg. Sociology and the Human Dimension. Cambridge: Cambridge University Press, 1982.

Brink, David O. Moral Realism and the Foundations of Ethics. Cambridge: Cambridge University Press, 1989.

Broad, C. D. Five Types of Ethical Theory. London: Kegan Paul, Trench, Trubner and Co., 1930.

Brooks, William Keith. The Foundations of Zoology. New York: Columbia University Press, 1899.

Brown, Alan Willard. The Metaphysical Society: Victorian Minds in Crisis, 1869–1880. New York: Columbia University Press, 1947.

Bruening, William H. The IS-OUGHT Problem: Its History, Analysis, and Dissolution. Washington, D.C.: University Press of America, 1978.

Bunge, Mario. Scientific Materialism. Dordrecht: D. Reidel, 1981.

Burkhardt, Frederick, and Sydney Smith, eds. The Correspondence of Charles Darwin. Cambridge: Cambridge University Press, 1985– .

Burrow, J. W. Evolution and Society: A Study in Victorian Social Theory. Cambridge: Cambridge University Press, 1966.

Burtt, E. A. "Present-Day Tendencies in Ethical Theory." *International Journal of Ethics* 31, no. 4 (1921): 432–438.

Bynum, William Frederick. "Time's Noblest Offspring: The Problem of Man in the British Natural Historical Sciences, 1800–1863." Ph.D. dissertation, Cambridge University, 1974.

Cacoullos, Ann R. Thomas Hill Green: Philosopher of Rights. New York: Twayne, 1974.

Calderwood, Henry. "Animal Ethics as Described by Herbert Spencer." *Philosophical Review* 1 (1892): 241–252.

———. "Ethical Aspects of the Theory of Development." *Contemporary Review* 31 (1877): 123–132.

———. Evolution and Man's Place in Nature. 2d ed. London: Macmillan, 1896. (First edition, 1893.)

Callahan, Daniel, and H. Tristram Engelhardt, Jr., eds. The Roots of Ethics: Science, Religion, and Values. New York: Plenum Press, 1981.

Cambridge University Examination Papers. Cambridge: Cambridge University Press, 1899–1987.

Campbell, Donald T. "On the Conflicts Between Biological and Social Evolution and Between Psychology and Moral Tradition." American Psychologist 30 (1975): 1103–1126.

———. Scientific Inquiry and the Social Sciences. San Francisco: Jossey-Bass, 1981.

———. "Social Morality Norms as Evidence of Conflict Between Biological Human Nature and Social Systems Requirements." In Gunther S. Stent, ed., Morality as a Biological Phenomenon: The Presuppositions of Sociobiology. Berkeley, Los Angeles, and London: University of California Press, 1980.

Campbell, T. D. Adam Smith's Science of Morals. London: George Allen and Unwin, 1971.

Caplan, Arthur, ed. The Sociobiology Debate: Readings on Ethical and Scientific Issues. New York: Harper and Row, 1978.

Caplan, Arthur, and Bruce Jennings, eds. Darwin, Marx and Freud: Their Influence on Moral Theory. New York: Plenum Press, 1984.

Carneino, Robert, ed. The Evolution of Society: Selections from Herbert Spencer's "Principles of Sociology." Chicago: University of Chicago Press, 1967.

Carr-Saunders, A. M. The Biological Basis of Human Nature. L. T. Hobhouse Memorial Trust Lectures, no. 12. London: Oxford University Press, 1942.

———. The Population Problem: A Study of Human Evolution. Oxford: Oxford University Press, 1922.

Cashdollar, Charles D. The Transformation of Theology, 1830–1890: Positivism and Protestant Thought in Britain and America. Princeton: Princeton University Press, 1989.

Cavalli-Sforza, L. L., and M. W. Feldman. Cultural Transmission and Evolution: A Quantitative Approach. Princeton: Princeton University Press, 1981.

Cela-Conde, Camilo J. "The Challenge of Evolutionary Ethics." Biology and Philosophy 1, no. 3 (1986): 293–297.

———. "Nature and Reason in the Darwinian Theory of Moral Sense." History and Philosophy of the Life Sciences 6 (1984): 3–24.

———. On Genes, Gods and Tyrants. Dordrecht: D. Reidel, 1987.

Chadwick, [William] Owen. The Secularization of the European Mind in the Nineteenth Century. Cambridge: Cambridge University Press, 1975.

———. The Victorian Church. 2 vols. London: Adam and Charles Black, 1966–1970.

Chagnon, Napoleon A., and William Irons, eds. Evolutionary Biology

and Human Social Behavior: An Anthropological Perspective. North
   Scituate: Duxbury Press, 1979.
Chambliss, J. J. "Natural Selection and Utilitarian Ethics in Chauncey
   Wright." American Quarterly 12 (1960): 144–159.
Clark, John Spencer. The Life and Letters of John Fiske. 2 vols. Boston:
   Houghton Mifflin, 1917.
Clifford, William Kingdon. The Common Sense of the Exact Sciences.
   Edited by Karl Pearson. New edition edited by James Newman. New
   York: Alfred A. Knopf, 1946.
———. Lectures and Essays by the Late William Kingdon Clifford. 2 vols.
   Edited by Leslie Stephen and Frederick Pollock. London: Macmillan,
   1879.
Cobbe, Frances Power. Darwinism in Morals, and Other Essays. London:
   Williams and Norgate, 1872.
Cockshut, A. O. J. The Unbelievers: English Agnostic Thought, 1840–
   1890. New York: New York University Press, 1966.
Coker, F. W. Organismic Theories of the State: Nineteenth-Century In-
   terpretations of the State as Organism or as Person. New York: Co-
   lumbia University Press, 1910.
Collingwood, R. G. Speculum Mentis: Or the Map of Knowledge. Ox-
   ford: Oxford University Press, 1924.
Collins, F. Howard. An Epitome of the Synthetic Philosophy. New York:
   D. Appleton, 1889.
Conklin, Edwin Grant. The Direction of Human Evolution. New York:
   Charles Scribner's Sons, 1921.
Copleston, Frederick. Modern Philosophy: Bentham to Russell. Vol. 8 of
   A History of Philosophy. Garden City: Image Books, 1967.
Courtney, W. L. "Modern Ethics." Edinburgh Review 157 (1883): 423–
   458.
Cowan, Ruth Schwartz. "Nature and Nurture: The Interplay of Biology
   and Politics in the Work of Francis Galton." Studies in History of Biol-
   ogy 1 (1977): 133–208.
Cravens, Hamilton. The Triumph of Evolution: American Scientists and
   the Hereditary-Environment Controversy 1900–1941. Philadelphia: Uni-
   versity of Pennsylvania Press, 1978.
Cronin, Helena. The Ant and the Peacock: Altruism and Sexual Selection
   from Darwin to Today. Cambridge: Cambridge University Press, 1991.
Crook, D. P. Benjamin Kidd: Portrait of a Social Darwinist. Cambridge:
   Cambridge University Press, 1984.
Crowther, M. A. Church Embattled: Religious Controversy in Mid-
   Victorian England. Newton Abbot: David and Charles, 1970.
Cunningham, G. Watts. The Idealistic Argument in Recent British and
   American Philosophy. New York: Century Co., 1933.
Cushing, James T., C. F. Delany, and Gary Gutting, eds. Science and

Reality: Recent Work in the Philosophy of Science. Essays in Honor of Ernan McMullin. Notre Dame: University of Notre Dame Press, 1984.

Darlington, C. D. The Evolution of Man and Society. London: George Allen and Unwin, 1969.

Darrah, William Culp. Powell of the Colorado. Princeton: Princeton University Press, 1951.

Darwell, Stephen, Allan Gibbard, and Peter Railton. "Toward Fin de siècle Ethics: Some Trends." Philosophical Review 101, no. 1 (1992): 115–189.

Darwin, Charles. The Descent of Man, and Selection in Relation to Sex. London: John Murray, 1871.

———. On the Origin of Species, Or the Preservation of Favoured Races in the Struggle for Life. Reprint of 1st ed. Cambridge: Harvard University Press, 1966.

Davies, John D. Phrenology, Fad and Science: A Nineteenth-Century American Crusade. Hamden: Archon Books, 1971.

Dawkins, Richard. "In Defense of Selfish Genes." Philosophy 56, no. 218 (1981): 556–573.

———. The Extended Phenotype: The Gene as the Unit of Selection. San Francisco: W. H. Freeman, 1982.

———. The Selfish Gene. Oxford: Oxford University Press, 1976.

Dawkins, W. Boyd. "The Descent of Man . . ." Edinburgh Review 134 (1871): 195–235.

De Giustino, David. Conquest of Mind: Phrenology and Victorian Social Thought. London: Croom Helm, 1975.

Degler, Carl N. In Search of Human Nature: The Decline and Revival of Darwinism in American Social Thought. Oxford: Oxford University Press, 1991.

de Laguna, T. "Stages of the Discussion of Evolutionary Ethics." Philosophical Review 14 (1905): 576–589.

Delius, Juan. "The Nature of Culture." In M. S. Dawkins, T. R. Halliday, and R. Dawkins, eds., The Tinbergen Legacy. London: Chapman and Hall, 1991. Pp. 75–99.

DeSalvo, Louise. Virginia Woolf: The Impact of Childhood Sexual Abuse on Her Life and Work. New York: Ballantine Books, 1989.

Desmond, Adrian. Archetypes and Ancestors: Palaeontology in Victorian London, 1850–1875. London: Blond and Briggs, 1982.

———. The Politics of Evolution: Morphology, Medicine, and Reform in Radical London. Chicago: University of Chicago Press, 1989.

———. "Robert E. Grant: The Social Predicament of a Pre-Darwinian Evolutionist." Journal of the History of Biology 17 (1984): 189–223.

Desmond, Adrian, and James Moore. The Life of a Tormented Evolutionist: Darwin. New York: Warner Books, 1991.

Dewey, John. "Evolution and Ethics." The Monist 8, no. 3 (1898): 321–341.

———. The Influence of Darwin on Philosophy and Other Essays in Contemporary Thought. New York: Henry Holt, 1910. Reprinted New York: Peter Smith, 1951.

———. Logic: The Theory of Inquiry. New York: Holt, Rinehart and Winston, 1938.

———. The Quest for Certainty: A Study of the Relation of Knowledge and Action. In John Dewey, The Later Works, 1925–1953. Vol. 4 (1929). Edited by Jo Ann Boydston. Carbondale: Southern Illinois University Press, 1984.

———. Theory of Valuation. Chicago: University of Chicago Press, 1939.

Dewey, John, and James H. Tufts. Ethics. New York: Henry Holt, 1908.

Dewey, John, et al. Creative Intelligence: Essays in the Pragmatic Attitude. New York: Holt, Rinehart and Winston, 1917. Reprinted New York: Octagon Books, 1970.

Digby, Anne, and Peter Searb. Children, School and Society in Nineteenth-Century England. London: Macmillan, 1981.

DiGregorio, Mario. T. H. Huxley's Place in Natural Science. New Haven: Yale University Press, 1984.

Dobzhansky, Theodosius. The Biological Basis of Human Freedom. New York: Columbia University Press, 1956.

———. The Biology of Ultimate Concern. London: Rapp and Whitney, 1969.

Docknell, D. W. "T. H. Huxley and the Meaning of 'Agnosticism.'" Theology 74, no. 616 (1971): 461–477.

Drummond, Henry. The Lowell Lectures on the Ascent of Man. London: Hodder and Stoughton, 1894.

———. Natural Law in the Spiritual World. London: Hodder and Stoughton, 1883.

Duncan, David, ed. Life and Letters of Herbert Spencer. 2 vols. New York: D. Appleton, 1968.

Durant, John R. "Evolution, Ideology and World View: Darwinian Religion in the Twentieth Century." In James Moore, ed., History, Humanity and Evolution: Essays for John C. Greene. Cambridge: Cambridge University Press, 1989. Pp. 355–373.

———. "Innate Character in Animals and Man: A Perspective on the Origins of Ethology." In Charles Webster, ed., Biology, Medicine, and Society, 1840–1940. Cambridge: Cambridge University Press, 1981. Pp. 157–192.

———. "Scientific Naturalism and Social Reform in the Thought of Alfred Russel Wallace." British Journal for the History of Science 12 (1979): 31–58.

Durant, John, ed. *Darwinism and Divinity: Essays on Evolution and Religious Belief.* Oxford: Basil Blackwell, 1985.

Ebling, F. J., ed. *Biology and Ethics.* New York: Academic Press, 1964.

Eisen, Sydney, and Bernard Lightman. *Victorian Science and Religion: A Bibliography with Emphasis on Evolution, Belief, and Unbelief, Comprised of Works Published from c. 1900–1975.* Hamden: Archon Books, 1984.

Ellegård, Alvar. "Darwin and the General Reader: The Reception of Darwin's Theory of Evolution in the British Periodical Press, 1859–1872." *Acta Universitatis Göthoburgensis* 64, no. 7 (1958).

———. "The Readership of the Periodical Press in Mid-Victorian Britain." *Acta Universitatis Göthoburgensis* 63, no. 3 (1957).

Ellery, John. *John Stuart Mill.* New York: Twayne Publishers, 1964.

Engel, A. J. *From Clergyman to Don: The Rise of the Academic Profession in Nineteenth-Century Oxford.* Oxford: Oxford University Press, 1983.

Essays and Reviews. 2d ed. London: John W. Parker and Son, 1860.

"Evolution and Man's Progress." *Daedalus* 90, no. 3 (1961).

Faber, Geoffrey. *Jowett: A Portrait with Background.* Cambridge: Harvard University Press, 1958.

Fairbrother, W. H. *The Philosophy of Thomas Hill Green.* London: Methuen, 1896.

Farber, Paul Lawrence. "The Transformation of Natural History in the Nineteenth Century." *Journal of the History of Biology* 15, no. 1 (1982): 145–152.

Fetzer, James H., ed. *Sociobiology and Epistemology.* Dordrecht: D. Reidel, 1985.

Fichman, Martin. *Alfred Russel Wallace.* Boston: Twayne, 1981.

———. "Ideological Factors in the Dissemination of Darwinism in England 1860–1900." In Everett Mendelsohn, ed., *Transformation and Tradition in the Sciences: Essays in Honor of I. Bernard Cohen.* Cambridge: Cambridge University Press, 1984. Pp. 471–485.

Fiske, John. *The Destiny of Man Viewed in the Light of His Origin.* Boston: Houghton Mifflin, 1884.

———. *Excursions of an Evolutionist.* Boston: Houghton Mifflin, 1899.

———. *The Idea of God as Affected by Modern Knowledge.* Boston: Houghton Mifflin, 1885.

———. *Outlines of Cosmic Philosophy, Based on the Doctrine of Evolution, with Criticisms on the Positive Philosophy.* 2 vols. Boston: Houghton Mifflin, 1874.

———. "The Progress from Brute to Man." *North American Review* 117 (1873): 251–319.

———. *Through Nature to God.* Boston: Houghton Mifflin, 1899.

———. *The Unseen World and Other Essays.* Boston: Houghton Mifflin, 1876.

Flew, Anthony. *Evolutionary Ethics.* London: Macmillan, 1967.

Foot, Philippa. *Virtues and Vices and Other Essays in Moral Philosophy.* Oxford: Basil Blackwell, 1978.

Forrest, D. W. *Francis Galton: The Life and Work of a Victorian Genius.* New York: Taplinger, 1974.

Fowle, T. W. "The Place of Conscience in Evolution." *Nineteenth Century* 4 (1878): 1–8.

Freeman, Derek. "The Evolutionary Theories of Charles Darwin and Herbert Spencer." *Current Anthropology* 15, no. 3 (1974): 211–221.

Galton, Francis. "Hereditary Talent and Character." *Macmillan's Magazine* 12 (1865): 157–166, 318–324.

———. *Inquiries into Human Faculty and Its Development.* London: Macmillan, 1883.

Gantz, Kenneth Franklin. "The Beginnings of Darwinian Ethics 1859–1871." *Studies in English, University of Texas Publication,* no. 3926 (1939): 180–209.

Gebhard, Geiger. "No Objective Values: A Critique of Ethical Intuitionism from an Evolutionary Point of View." *Biology and Philosophy* 7 (1992): 315–330.

Geddes, Patrick, and J. Arthur Thomson. *The Evolution of Sex.* London: Walter Scott, 1889.

George, Henry. *Progress and Poverty: An Inquiry into the Cause of Industrial Depressions and of Increase of Want with Increase of Wealth: The Remedy.* New York: D. Appleton, 1880.

Gerard, R. W. "A Biological Basis for Ethics." *Philosophy of Science* 9 (1942): 92–120.

Gewirth, Alan. "The Problem of Specificity in Evolutionary Ethics." *Biology and Philosophy* 1, no. 3 (1986): 297–305.

———. *Reason and Morality.* Chicago: University of Chicago Press, 1978.

Ginsberg, Morris. *Moral Progress.* Glasgow: Jackson and Son, 1944.

———. *Reason and Experience in Ethics.* London: Oxford University Press, 1956.

Goetzmann, William H., ed. *The American Hegelians: An Intellectual Episode in the History of Western America.* New York: Alfred A. Knopf, 1973.

Goldsmith, Timothy H. *The Biological Roots of Human Nature: Forging Links between Evolution and Behavior.* Oxford: Oxford University Press, 1991.

Gordon, Scott. "Darwin and Political Economy: The Connection Reconsidered." *Journal of the History of Biology* 22, no. 3 (1989): 437–459.

Gorer, Geoffrey. *Exploring English Character.* New York: Criterion Books, 1955.

Gouinlock, James. *John Dewey's Philosophy of Value*. New York: Humanities Press, 1972.

Gould, Stephen Jay. *Ever Since Darwin: Reflections in Natural History*. New York: W. W. Norton, 1977.

———. "On Replacing the Idea of Progress with an Operational Notion of Directionality." In Matthew H. Nitecki, ed., *Evolutionary Progress*. Chicago: University of Chicago Press, 1988. Pp. 319–336.

———. *Wonderful Life: The Burgess Shale and the Nature of History*. New York: W. W. Norton, 1989.

Grant, Alex. "On the Nature and Origin of Moral Ideas." *Fortnightly Review* 15 (1871): 363–373.

———. "Philosophy and Mr. Darwin." *Contemporary Review* 17 (1871): 274–281.

Green, Thomas Hill. *Prolegomena to Ethics*, ed. A. C. Bradley. Oxford: Oxford University Press, 1883.

Greene, John C. "Darwin as a Social Evolutionist." *Journal of the History of Biology* 8, no. 1 (1977): 1–27.

———. "The Interaction of Science and Worldview in Sir Julian Huxley's Evolutionary Biology." *Journal of the History of Biology* 23, no. 1 (1990): 39–55.

———. "Reflections on the Progress of Darwin Studies." *Journal of the History of Biology* 8, no. 2 (1975): 243–273.

———. *Science, Ideology, and World View: Essays in the History of Evolutionary Ideas*. Berkeley, Los Angeles, and London: University of California Press, 1981.

Greenwood, Davydd J. *The Taming of Evolution: The Persistence of Nonevolutionary Views in the Study of Humans*. Ithaca: Cornell University Press, 1984.

Gregory, Michael S., Anita Silvers, and Diane Sutch, eds. *Sociobiology and Human Nature*. San Francisco: Jossey-Bass, 1978.

Gruber, Howard, ed. *Darwin on Man: A Psychological Study of Scientific Creativity, Together with Darwin's Early and Unpublished Notebooks Transcribed and Annotated by Paul H. Barrett*. New York: Dutton, 1974.

Guyau, Marie Jean. *La morale Anglaise contemporaire: Morale de l'utilité et de l'évolution*. Paris: Librairie Germer Ballière, 1879.

Haeckel, Ernest. Natürliche Schöpfungsgeschichte. Berlin: Georg Reimer, 1868.

Haines, Valerie A. "Spencer, Darwin, and the Question of Reciprocal Influence." *Journal of the History of Biology* 24, no. 3 (1991): 409–431.

Haller, John S. *Outcasts of Evolution: Scientific Attitudes of Racial Inferiority, 1859–1900*. Urbana: University of Illinois Press, 1971.

Hamerton, P. G. "A Basis of Positive Morality." *Contemporary Review* 59 (1891): 537–545, 889–899.

Hamilton, Gail. *The Insuppressible Book: A Controversy Between Herbert Spencer and Frederic Harrison.* Boston: S. E. Cassino, 1885.

Hamilton, William. "The Genetical Theory of Social Behaviour." 2 pts. *Journal of Theoretical Biology* 7 (1964): 1–16, 17–52.

———. "Innate Social Aptitudes of Man: An Approach from Evolutionary Genetics." In Robin Fox, ed., *Biosocial Anthropology.* London: Malaby Press, 1975. Pp. 133–155.

Hampshire, Stuart. *Innocence and Experience.* Cambridge: Harvard University Press, 1989.

———. *Morality and Conflict.* Oxford: Basil Blackwell, 1983.

Hancock, Robert N. *Twentieth-Century Ethics.* New York: Columbia University Press, 1974.

Haraway, Donna Jeanne. *Crystals, Fabrics, and Fields: Metaphors of Organicism in Twentieth-Century Developmental Biology.* New Haven: Yale University Press, 1976.

Hardy, Alister. *Darwin and the Spirit of Man.* London: Collins, 1984.

Harrison, J. F. C. *The Early Victorians, 1832–1851.* New York: Praeger, 1971.

Harrod, R. F. *The Life of John Maynard Keynes.* New York: Harcourt, Brace, 1951.

Hauerwas, Stanley, and Alasdair MacIntyre, eds. *Revisions: Changing Perspectives in Moral Philosophy.* Notre Dame: University of Notre Dame, 1983.

Haycraft, John Berry. *Darwinism and Race Progress.* London: Swan Sonnerschein, 1895.

Hayek, F. A. *The Counter-Revolution of Science: Studies on the Abuse of Reason.* New York: Free Press, 1955.

Hays, Samuel. *The Response to Industrialization 1885–1914.* Chicago: University of Chicago Press, 1957.

Hefland, Michael. "T. H. Huxley's 'Evolution and Ethics': The Politics of Evolution and the Evolution of Politics." *Victorian Studies* 20, no. 2 (1977): 159–177.

Henderson, G. C. "Natural and Rational Selection." *International Journal of Ethics* 24, no. 2 (1914): 127–147.

Herbert, Sandra. "The Place of Man in the Development of Darwin's Theory of Transmutation." 2 pts. *Journal of the History of Biology* 7, no. 2 (1974): 217–258, and 10, no. 2 (1977): 155–227.

Heyck, T. W. *The Transformation of Intellectual Life in Victorian England.* New York: St. Martin's, 1982.

Hinchliff, Peter. *Benjamin Jowett and the Christian Religion.* Oxford: Oxford University Press, 1987.

Hinsley, Curtis M., Jr. *Savages and Scientists: The Smithsonian Institution and the Development of American Anthropology, 1846–1910.* Washington, D.C.: Smithsonian Institution Press, 1981.

Hobhouse, L. T. *Mind in Evolution*. London: Macmillan, 1915.

Hofstadter, Richard. *Social Darwinism in American Thought*. Philadelphia: University of Pennsylvania Press, 1944.

Holbach, Henry. "Sir Alexander Grant on the 'Nature and Origin of Moral Ideas.'" *Contemporary Review* 17 (1871): 299–306.

Holmes, Samuel Jackson. *Life and Morals*. New York: Macmillan, 1948.

Houghton, Walter, et al., eds. *The Wellesley Index to Victorian Periodicals 1824–1900*. 5 vols. Toronto: University of Toronto Press, 1966–1989.

Hudson, William Donald. *Modern Moral Philosophy*. 2d ed. London: Macmillan, 1983.

Hudson, William Donald, ed. *The IS/OUGHT Question: A Collection of Papers on the Central Problems in Moral Philosophy*. London: Macmillan, 1969.

Hudson, William Henry. *An Introduction to the Philosophy of Herbert Spencer*. London: Chapman and Hall, 1897. Reprinted, New York: Haskell House Publishers, 1974.

Hughes, Henry Stuart. *Consciousness and Society: The Reorientation of Social Thought 1890–1930*. New York: Alfred A. Knopf, 1958.

Hughes, William. "Richards's Defense of Evolutionary Ethics." *Biology and Philosophy* 1, no. 3 (1986): 306–315.

Hume, David. *A Treatise of Human Nature*. Edited by L. A. Selby-Bigge. Oxford: Oxford University Press, 1960. (First edition, 1739.)

Hutchinson, Woods. *Civilization and Health*. Boston: Houghton Mifflin, 1914.

———. *Common Diseases*. Boston: Houghton Mifflin, 1913.

———. *The Gospel According to Darwin*. Chicago: Open Court, 1898.

———. *A Handbook of Health*. Boston: Houghton Mifflin, 1911.

———. *Instinct and Health*. New York: Dodd and Mead, 1909.

———. *Preventable Diseases*. Boston and New York: Houghton Mifflin, 1909.

Hutton, Frederick Wollaston. *Darwinism and Lamarckism: Old and New, Four Lectures*. London: Duckworth, 1899.

Hutton, Richard. "The Natural History of Morals." *Westminster Review* n.s., 36 (1869): 494–531.

Huxley, Julian. *Essays of a Biologist*. London: Chatto and Windus, 1923.

———. *Essays of a Humanist*. London: Chatto and Windus, 1964.

———. *Evolutionary Ethics*. Oxford: Oxford University Press, 1943.

———. *Evolution: The Modern Synthesis*. New York: Harper and Brothers, 1943.

———. *Memories*. London: George Allen and Unwin, 1970.

———. *New Bottles for New Wine*. London: Chatto and Windus, 1957.

———. *Religion without Revelation*. London: Ernest Benn, 1927.

————. *Science, Religion, and Human Nature*. Conway Memorial Lecture. London: Watts, 1930.

————. *The Uniqueness of Man*. London: Chatto and Windus, 1941.

————. What Dare I Think? The Challenge of Modern Science to Human Action and Belief. London: Chatto and Windus, 1933. (First edition, 1931.)

Huxley, Julian, ed. *The Humanist Frame: The Modern Humanist Vision of Life*. New York: Harper and Brothers, 1961.

Huxley, Juliette. *Leaves of the Tulip Tree*. London: John Murray, 1986.

Huxley, Leonard. *Life and Letters of Thomas Henry Huxley*. 2 vols. New York: D. Appleton, 1900.

Huxley, Thomas Henry. *Collected Essays*. London: Macmillan, 1893–1894.

————. *Evidence as to Man's Place in Nature*. London: Williams and Norgate, 1863.

————. "On the Physical Basis of Life." *Fortnightly Review* 5, n.s. (1868): 129–145.

————. "The Struggle for Existence." *Nineteenth Century* 23 (1888): 161–180.

Huxley, Thomas Henry, and Julian Huxley. *Touchstone for Ethics 1893–1943*. New York: Harper and Brothers, 1947.

Irvine, William. *Apes, Angels, and Victorians: Darwin, Huxley, and Evolution*. New York: McGraw-Hill, 1955.

————. *Walter Bagehot*. London: Longmans, Green, 1939.

James, William. *Collected Essays and Reviews*. London: Longmans, Green, 1920.

————. *Pragmatism: A New Name for Some Old Ways of Thinking, Together with Four Related Essays Selected from the Meaning of Truth*. London: Longmans, Green, 1907.

Jensen, J. Vernon. "The X Club: Fraternity of Victorian Scientists." *British Journal for the History of Science* 5, no. 17 (1970): 63–72.

————. *Thomas Henry Huxley: Communicating for Science*. Newark: University of Delaware Press, 1991.

Jensen, W. J., and R. Harré, eds. *The Philosophy of Evolution*. New York: St. Martin's, 1981.

Johnston, Paul. *Wittgenstein and Moral Philosophy*. London: Routledge, 1989.

Jones, Greta. *Social Darwinism and English Thought*. Atlantic Highlands, N.J.: Humanities Press, 1980.

Kary, Carla. "Sociobiology and the Redemption of Normative Ethics." *The Monist* 67, no. 2 (1984): 161–166.

Kaye, Howard L. *The Social Meaning of Modern Biology: From Social Darwinism to Sociobiology*. New Haven: Yale University Press, 1986.

Keith, Sir Arthur. *Evolution and Ethics.* New York: Putnam, 1946.

Keller, Evelyn Fox. "Nature, Nurture, and the Human Genome Project." In Daniel J. Kevles and Leroy Hood, eds., *The Code of Codes: Scientific and Social Issues in the Human Genome Project.* Cambridge: Harvard University Press, 1992. Pp. 281–299.

Kellogg, Vernon. *Darwinism Today: A Discussion of Present-Day Scientific Criticism of the Darwinian Selection Theories, Together with a Brief Account of the Principal Other Proposed Auxiliary and Alternative Theories of Species-Forming.* New York: Henry Holt, 1907.

Kelly, Alfred. *The Descent of Darwin: The Popularization of Darwinism in Germany, 1860–1914.* Chapel Hill: University of North Carolina Press, 1981.

Kennedy, James G. *Herbert Spencer.* Boston: Twayne Publishers, 1978.

Kevles, Daniel J., and Leroy Hood, eds. *The Code of Codes: Scientific and Social Issues in the Human Genome Project.* Cambridge: Harvard University Press, 1992.

Keynes, John Maynard. Essays in Biography. London: Macmillan, 1933.

Keynes, Milo, and Harrison G. Ainsworth, eds. *Evolutionary Studies: A Centenary Celebration of the Life of Julian Huxley.* Proceedings of the Twenty-fourth Annual Symposium of the Eugenics Society, 1987. London: Macmillan, 1989.

Kidd, Benjamin. *A Philosopher with Nature.* London: Methuen, 1921.

———. *The Science of Power.* New York and London: Putnam's Sons, 1918.

———. *Social Evolution.* New York: Macmillan, 1894.

King's College Sociobiology Group, Cambridge, ed. *Current Problems in Sociobiology.* Cambridge: Cambridge University Press, 1982.

Kingsland, Sharon. "Toward a Natural History of the Human Psyche: Charles Manning Child, Charles Judson Herrick, and the Dynamic View of the Individual at the University of Chicago." In Keith R. Benson, Jane Maienschein, and Ronald Rainger, eds., *The Expansion of American Biology.* New Brunswick: Rutgers University Press, 1991. Pp. 195–230.

Kitcher, Philip. *Vaulting Ambition: Sociobiology and the Quest for Human Nature.* Cambridge: MIT Press, 1985.

Knight, William. "Ethical Philosophy and Evolution." *Nineteenth Century* 4 (1878): 432–456.

Knorr, Karin, H. Strasser, and H. Zilian, eds. *Determinants and Controls of Scientific Development.* Dordrecht: D. Reidel, 1975.

Knorr-Cetina, Karin D. *The Manufacture of Knowledge: An Essay on the Constructivist and Contextual Nature of Science.* Oxford: Pergamon Press, 1981.

Kottler, Malcolm Jay. "Alfred Russel Wallace, the Origin of Man, and Spiritualism." *Isis* 65, no. 227 (1974): 145–192.

Kurtz, Paul. *Philosophical Essays in Pragmatic Naturalism*. Buffalo: Prometheus Books, 1990.

Kurtz, Paul, ed. *The Humanist Alternative: Some Definitions of Humanism*. London: Pemberton, 1973.

Lamont, W. D. *Introduction to Green's Moral Philosophy*. London: George Allen and Unwin, 1934.

Lankester, Sir Ray. *Great and Small Things*. New York: Macmillan, 1923.

———. *Science from an Easy Chair: A Second Series*. New York: Henry Holt, 1913.

———. *Secrets of Earth and Sea*. London: Methuen, 1920.

Larson, Edward. *Trial and Error: The American Controversy Over Creation and Evolution*. Oxford: Oxford University Press, 1985.

La Vergata, Antonello. "Images of Darwin: A Historiographic Overview." In David Kohn, ed., *The Darwinian Heritage*. Princeton: Princeton University Press, 1985. Pp. 901–972.

Leake, Chauncey. "An American Opinion." In C. H. Waddington, *Science and Ethics*. London: George Allen and Unwin, 1942. P. 133.

Leake, Chauncey, and Patrick Romanell. *Can We Agree? A Scientist and a Philosopher Argue about Ethics*. Austin: University of Texas Press, 1950.

Lester, Jacob. "John Fiske's Philosophy of Science: The Union of Science and Religion Through the Principle of Evolution." Ph.D. dissertation, Oregon State University, 1979.

LeSueur, William D. "A Vindication of Scientific Ethics." *Popular Science Monthly* 17 (1880): 324–337.

Levy, Paul. *G. E. Moore and the Cambridge Apostles*. New York: Holt, Rinehart and Winston, 1979.

Lewontin, R. C., Steven Rose, and Leon Kamin. *Not in Our Genes: Biology, Ideology, and Human Nature*. New York: Pantheon Books, 1984.

Lightman, Bernard. *The Origins of Agnosticism: Victorian Unbelief and the Limits of Knowledge*. Baltimore: Johns Hopkins University Press, 1987.

Lilly, William Samuel. *On Right and Wrong*. London: Chapman and Hall, 1890.

———. "Our Great Philosopher." 2 pts. *Contemporary Review* 55 (1889): 752–770, and 56 (1889): 586–609.

Lippmann, Walter. *A Preface to Morals*. New York: Macmillan, 1929.

Livingstone, David N. "Darwin and Calvinism: The Belfast-Princeton Connection." *Isis* 83 (1992): 408–428.

———. *Darwin's Forgotten Defenders: The Encounter Between Evangelical Theology and Evolutionary Thought*. Edinburgh: Scottish Academic Press, 1987.

Lumsden, Charles J., and Edward O. Wilson. *Genes, Mind and Culture: The Coevolutionary Process.* Cambridge: Harvard University Press, 1981.

———. *Promethean Fire: Reflections on the Origin of Mind.* Cambridge: Harvard University Press, 1983.

Lytle, Guy Fitch, and Stephen Orgel, eds. *Patronage in the Renaissance.* Princeton: Princeton University Press, 1981.

McGinn, Colin. "Evolution, Animals, and the Basis of Morality." *Inquiry* 22 (1979): 81–99.

Machin, Alfred. *The Ascent of Man by Means of Natural Selection.* London: Longmans, Green, 1925.

———. *Darwin's Theory Applied to Mankind.* Foreword by Sir Arthur Keith. London: Longmans, Green, 1937.

MacIntosh, Robert. *From Comte to Benjamin Kidd: The Appeal to Biology or Evolution for Human Guidance.* New York: Macmillan, 1899.

MacIntyre, Alasdair. *After Virtue.* Notre Dame: University of Notre Dame Press, 1981.

———. *Against the Self-Images of the Age: Essays on Ideology and Philosophy.* Notre Dame: University of Notre Dame Press, 1984. (First edition, 1971.)

———. "Moral Philosophy: What Next?" In Stanley Hauerwas and Alasdair MacIntyre, eds., *Revisions: Changing Perspectives in Moral Philosophy.* Notre Dame: University of Notre Dame Press, 1983. Pp. 1–15.

———. *Secularization and Moral Change.* Oxford: Oxford University Press, 1967.

———. *A Short History of Ethics.* New York: Macmillan, 1966.

———. *Three Rival Versions of Moral Enquiry: Encyclopedia, Genealogy, and Tradition.* Notre Dame: University of Notre Dame Press, 1990.

———. *Whose Justice? Which Rationality?* Notre Dame: University of Notre Dame Press, 1988.

Mackie, J. L. *Ethics: Inventing Right and Wrong.* London: Penguin Books, 1977.

———. "Genes and Egoism." *Philosophy* 56, no. 218 (1981): 553–555.

———. "The Law of the Jungle: Moral Alternatives and Principles of Evolution." *Philosophy* 53, no. 206 (1978): 455–464.

———. *Persons and Values: Selected Papers. Vol.* 2. Edited by Joan Mackie and Penelope Mackie. Oxford: Oxford University Press, 1985.

Mackintosh, James. *Dissertation on the Progress of Ethical Philosophy, Chiefly during the Seventeenth and Eighteenth Centuries.* Edinburgh: Adam and Charles Black, 1836.

MacLeod, Roy. "The X-Club: A Social Network of Science in Late-

Victorian England." *Notes and Records of the Royal Society of London* 24 (1969): 305–322.

Madden, Edward H. *Chauncey Wright and the Foundations of Pragmatism.* Seattle: University of Washington Press, 1963.

Maitland, Frederick William. *The Life and Letters of Leslie Stephen.* London: Duckworth, 1906.

Manier, Edward. *The Young Darwin and His Cultural Circle.* Dordrecht: D. Reidel, 1978.

Marcell, David W. *Progress and Pragmatism: James, Dewey, Beard, and the American Idea of Progress.* Westport, Conn.: Greenwood Press, 1974.

Marchant, James, ed. *Alfred Russel Wallace: Letters and Reminiscences.* London: Cassel, 1916.

Marsh, P. T. *The Victorian Church in Decline: Archbishop Tait and the Church of England 1868–1882.* London: Routledge and Kegan Paul, 1969.

Martineau, Harriet. *Autobiography.* Edited by Maria Weston Chapman. Boston: James R. Osgood, 1878.

———. *How to Observe: Morals and Manners.* London: C. Knight, 1838.

Mathews, Robert. "Evolutionary Ethics." *Popular Science Monthly* 44 (1893): 192–195.

Mayr, Ernst, and William Provine, eds. *The Evolutionary Synthesis: Perspectives on the Unification of Biology.* Cambridge: Harvard University Press, 1980.

Mead, George. "The Philosophical Basis of Ethics." *International Journal of Ethics* 18, no. 3 (1908): 311–323.

———. "Scientific Method and the Moral Sciences." *International Journal of Ethics* 33, no. 3 (1923): 229–247.

Midgley, Mary. *Beast and Man: The Roots of Human Nature.* Ithaca: Cornell University Press, 1978.

———. *Evolution as a Religion: Strange Hopes and Stranger Fears.* London: Methuen, 1985.

———. "Gene-Juggling." *Philosophy* 54, no. 210 (1979): 439–458.

———. *Heart and Mind: The Varieties of Moral Experience.* New York: St. Martin's, 1981.

———. "Rival Fatalisms: The Hollowness of the Sociobiology Debate." In Ashley Montagu, ed., *Sociobiology Examined.* New York: Oxford University Press, 1980. Pp. 15–38.

Mill, John Stuart. *Collected Works of John Stuart Mill.* Toronto: University of Toronto Press, 1969.

Miller, Hugh. *Progress and Decline: The Grouper in Evolution.* Los Angeles: Ward Ritchie Press, 1963.

[Millin, George]. *Evil and Evolution: An Attempt to Turn the Light of*

Modern Science on to the Ancient Mystery of Evil. London: Macmillan, 1896.

Mitman, Gregg. "Evolution as Gospel: William Patten, the Language of Democracy, and the Great War." *Isis* 81, no. 308 (1990): 446–463.

———. "From Population to Society: The Cooperative Metaphors of W. C. Allee and A. E. Emerson." *Journal of the History of Biology* 21, no. 2 (1988): 173–194.

———. *The State of Nature: Ecology, Community, and American Social Thought, 1900–1950*. Chicago: University of Chicago Press, 1992.

Mivart, Jackson St. George. *On the Genesis of Species*. 2d ed. London: Macmillan, 1871.

Montagu, Ashley, ed. *Sociobiology Examined*. New York: Oxford University Press, 1980.

Moore, Edward C. *American Pragmatism: Peirce, James, and Dewey*. New York: Columbia University Press, 1961.

Moore, George Edward. *Ethics*. Oxford: Oxford University Press, 1965. (First edition, 1912.)

———. *Principia Ethica*. Cambridge: Cambridge University Press, 1978. (First edition, 1903.)

Moore, James. "Herbert Spencer's Henchmen: The Evolution of Protestant Liberals in Late Nineteenth-Century America." In John Durant, ed., *Darwinism and Divinity: Essays on Evolution and Religious Belief*. Oxford: Basil Blackwell, 1985. Pp. 76–100.

———. *The Post-Darwinian Controversies: A Study of the Protestant Struggle to Come to Terms with Darwin in Great Britain and America 1870–1900*. Cambridge: Cambridge University Press, 1979.

Morgan, Conway Lloyd. *The Emergence of Novelty*. New York: Henry Holt, n.d.

———. *Emergent Evolution*. Gifford Lectures of 1922. New York: Henry Holt, 1923.

———. *Habit and Instinct*. London: E. Arnold, 1896.

———. *Instinct and Experience*. New York: Macmillan, 1912.

———. *Life, Mind, and Spirit, Being the Second Course of the Gifford Lectures*. New York: Henry Holt, 1925.

Morris, Charles. *The Pragmatic Movement in American Philosophy*. New York: George Braziller, 1970.

Muirhead, John. *The Platonic Tradition in Anglo-Saxon Philosophy: Studies in the History of Idealism in England and America*. London: George Allen and Unwin, 1931.

Mulkay, Michael. *Science and the Sociology of Knowledge*. London: George Allen and Unwin, 1979.

Munitz, Milton. *Contemporary Analytical Philosophy*. New York: Macmillan, 1981.

Murphy, Howard. "The Ethical Revolt Against Christian Orthodoxy in

Early Victorian England." *American Historical Review* 60, no. 4 (1955): 800–817.

Murphy, Jeffrie G. *Evolution, Morality, and the Meaning of Life.* Totowa: Rowman and Littlefield, 1982.

Myers, Gerald E. *William James: His Life and Thought.* New Haven: Yale University Press, 1986.

Needham, Joseph. *The Great Amphibium: Four Lectures on the Position of Religion in a World Dominated by Science.* London: Student Christian Movement Press, 1931.

————. [published under pseudonym, Henry Holorenshaw] "The Making of an Honorary Taoist." In Mikuláš Teich and Robert Young, eds., *Changing Perspectives in the History of Science: Essays in Honour of Joseph Needham.* London: Heinemann, 1973. Pp. 1–20.

————. *The Sceptical Biologist.* London: Chatto and Windus, 1929.

————. *Time: The Refreshing River, Essays and Addresses, 1932–1942.* London: George Allen and Unwin, 1943.

Nettleship, R. L., ed. *Works of Thomas Hill Green.* London: Longmans, Green, 1906.

Nisbet, J. F. *Marriage and Heredity: A View of Psychological Evolution.* London: Ward and Downey, 1889.

Nisbet, Robert. *History of the Idea of Progress.* New York: Basic Books, 1980.

Nitecki, Matthew H., ed. *Evolutionary Progress.* Chicago: University of Chicago Press, 1988.

Norton, Charles Elliot, ed. *Philosophical Discussions by Chauncey Wright with a Biographical Sketch of the Author.* New York: Henry Holt, 1877.

O'Donnell, John M. *The Origins of Behaviorism: American Psychology, 1870–1920.* New York: New York University Press, 1985.

Oldroyd, David R. *Darwinian Impacts, an Introduction to the Darwin Revolution.* Atlantic Highlands, N.J.: Humanities Press, 1980.

————. "How Did Darwin Arrive at His Theory? The Secondary Literature to 1982." *History of Science* 22 (1984): 325–374.

Oldroyd, David, and Ian Langham, eds. *The Wider Domain of Evolutionary Thought.* Dordrecht: D. Reidel, 1983.

Oppenheim, Janet. *The Other World: Spiritualism and Psychical Research in England, 1850–1914.* Cambridge: Cambridge University Press, 1985.

Otto, Rudolf. *The Idea of the Holy: An Inquiry into the Non-Rational Factor in the Idea of the Divine and Its Relation to the Rational.* Oxford: Oxford University Press, 1923.

Pannill, H. Burnell. *The Religious Faith of John Fiske.* Durham: Duke University Press, 1957.

Paradis, James. *T. H. Huxley: Man's Place in Nature.* Lincoln: University of Nebraska Press, 1978.

Paradis, James, and Thomas Postlewait, eds. *Victorian Science and Victorian Values: Literary Perspectives.* MLF Symposium, 1978. Annals of the New York Academy of Sciences 360 (1981).

Paradis, James, and George Williams, eds. *Evolution and Ethics: T. H. Huxley's Evolution and Ethics with New Essays on Its Victorian and Sociobiological Context.* Princeton: Princeton University Press, 1989.

Patrick, James. *The Magdalen Metaphysicals: Idealism and Orthodoxy at Oxford, 1901–1945.* Macon: Mercer University Press, 1985.

Patten, William. *The Grand Strategy.* Boston: Richard G. Badger, 1920.

Paxton, Nancy. *George Eliot and Herbert Spencer.* Princeton: Princeton University Press, 1991.

Peacocke, A. R., ed. *The Science and Theology in the Twentieth Century.* Stocksfield: Oriel Press, 1981.

Peel, J. D. Y. *Herbert Spencer: The Evolution of a Sociologist.* New York: Basic Books, 1971.

———. "Spencer and the Neo-Evolutionists." *Sociology* 3 (1969): 173–191.

Pepper, Stephen. *Ethics.* New York: Appleton-Century-Crofts, 1960.

Perkin, Harold. *The Origins of Modern English Society 1780–1880.* London: Kegan Paul, 1969.

Perry, Ralph Barton. *Philosophy of the Recent Past: An Outline of European and American Philosophy Since 1860.* London: Charles Scribner's Sons, 1927.

———. *The Thought and Character of William James.* Boston: Little, Brown, 1935.

Pickens, Donald K. *Eugenics and the Progressives.* Nashville: Vanderbilt University Press, 1968.

Pollock, Frederick. "Evolution and Ethics." *Mind* 1, no. 3 (1876): 334–345.

Poulton, E. B. "Alfred Russel Wallace, 1823–1913." *Proceedings of the Royal Society of London,* Series B, 95 (1923–1924) i–xxxv.

Provine, William. *The Origins of Theoretical Population Genetics.* Chicago: University of Chicago Press, 1971.

Pugh, George Edgin. *The Biological Origin of Human Values.* New York: Basic Books, 1977.

Putnam, Hilary. *Mind, Language and Reality: Philosophical Papers.* Vol. 2. Cambridge: Harvard University Press, 1975.

Quillian, William, Jr. *The Moral Theory of Evolutionary Naturalism.* New Haven: Yale University Press, 1945.

Quinton, Anthony. "Ethics and the Theory of Evolution." In I. T. Ramsey, ed., *Biology and Personality: Frontier Problems in Science, Philosophy and Religion.* Oxford: Basil Blackwell, 1965. Pp. 107–131.

———. *Utilitarian Ethics.* New York: St. Martin's, 1973.

Rainger, Ronald, Keith Benson, and Jane Maienschein, eds. *The Ameri-*

can Development of Biology. Philadelphia: University of Pennsylvania Press, 1988.

Raven, Charles E. *The Creator Spirit: A Survey of Christian Doctrine in the Light of Biology, Psychology and Mysticism*. London: Martin Hopkinson, 1927.

——. *Science, Religion, and the Future*. Cambridge: Cambridge University Press, 1943.

Rawls, John. *A Theory of Justice*. Cambridge: Harvard University Press, 1971.

Regan, Tom. *Bloomsbury's Prophet: G. E. Moore and the Development of His Moral Philosophy*. Philadelphia: Temple University Press, 1986.

Resek, Carl. *Lewis Henry Morgan: American Scholar*. Chicago: University of Chicago Press, 1960.

Rice, Philip Blair. "Objectivity in Value Judgments." *Journal of Philosophy* 4, no. 1 (1943): 5–14.

Richards, Evelleen. "Huxley and Woman's Place in Science: The 'Woman Question' and Control of Victorian Anthropology." In James Moore, ed., *History, Humanity and Evolution: Essays for John C. Greene*. Cambridge: Cambridge University Press, 1989. Pp. 253–284.

——. "The 'Moral Anatomy' of Robert Knox: The Interplay Between Biological and Social Thought in Victorian Scientific Naturalism." *Journal of the History of Biology* 22, no. 3 (1989): 373–436.

Richards, Robert J. *Darwin and the Emergence of Evolutionary Theories of Mind and Behavior*. Chicago: University of Chicago Press, 1987.

——. "A Defense of Evolutionary Ethics." *Biology and Philosophy* 1, no. 3 (1986): 265–293.

——. "Dutch Objections to Evolutionary Ethics." *Biology and Philosophy* 4 (1989): 331–343.

——. "Justification Through Biological Faith: A Rejoinder." *Biology and Philosophy* 1, no. 3 (1986): 337–354.

Richter, Melvin. *The Politics of Conscience: T. H. Green and His Age*. London: Weidenfeld and Nicolson, 1964.

Riedl, Rupert. *Biology of Knowledge: The Evolutionary Basis of Reason*. New York: John Wiley and Sons, 1984.

Ritchie, David G. *Darwinism and Politics: With Two Additional Essays on Human Evolution*. 3d ed. London: Swan Sonnenschein, 1895. (First edition, 1889.)

——. "Natural Selection and the Spiritual World." *Westminster Review* 133 (1890): 459–469.

Ritter, William Emerson. *Charles Darwin and the Golden Rule*. Edited by Edna Watson Bailey. Washington, D.C.: Science Service, 1954.

Romanes, George John. *Mental Evolution in Man: Origin of Human Faculty*. New York: D. Appleton, 1893.

Rorty, Richard. *Consequences of Pragmatism: Essays, 1972–1980.* Minneapolis: University of Minnesota Press, 1982.

———. *Contingency, Irony, and Solidarity.* Cambridge: Cambridge University Press, 1989.

———. *Philosophy and the Mirror of Nature.* Princeton: Princeton University Press, 1979.

Rosenberg, Charles. *No Other Gods: On Science and American Social Thought.* Baltimore: Johns Hopkins University Press, 1976.

Roth, John K. *Freedom and the Moral Life: The Ethics of William James.* Philadelphia: Westminster Press, 1969.

Rothblatt, Sheldon. *The Revolution of the Dons: Cambridge and Society in Victorian England.* London: Faber and Faber, 1968.

Rottschaefer, William. "Evolutionary Naturalistic Justification of Morality: A Matter of Faith and Works." *Biology and Philosophy* 6 (1991): 341–349.

Roubiczek, Paul. *Ethical Values in the Age of Science.* Cambridge: Cambridge University Press, 1969.

Royce, Josiah. "Report on the Recent Literature of Ethics and Related Topics in America." *International Journal of Ethics* 3, no. 4 (1893): 527–541.

Ruse, Michael. *The Darwinian Revolution.* Chicago: University of Chicago Press, 1979.

———. "The Morality of the Gene." *The Monist* 67, no. 2 (1984): 167–199.

———. *Sociobiology: Sense or Nonsense?* Dordrecht: D. Reidel, 1979.

———. *Taking Darwin Seriously: A Naturalistic Approach to Philosophy.* Oxford: Basil Blackwell, 1986.

Ryan, Alan. *The Philosophy of John Stuart Mill.* Atlantic Highlands: Humanities Press, 1970.

Sahlins, Marshall. *The Use and Abuse of Biology: An Anthropological Critique of Sociobiology.* Ann Arbor: University of Michigan Press, 1976.

Savage, Minot J. *The Morals of Evolution.* Boston: George H. Ellis, 1880.

———. *The Religion of Evolution.* Boston: Lockwood, Brooks, 1876.

Schilcher, Florian von, and Neil Tennant. *Philosophy, Evolution and Human Nature.* London: Routledge and Kegan Paul, 1984.

Schilpp, Paul Arthur, ed. *The Philosophy of John Dewey.* 2d ed. New York: Tudor, 1951. (First edition, 1939.)

Schneewind, J. B. *Sidgwick's Ethics and Victorian Moral Philosophy.* Oxford: Oxford University Press, 1977.

Schoeck, Helmut, and James W. Wiggins, eds. *Scientism and Values.* Princeton: D. Van Nostrand, 1960.

Schurman, Jacob Gould. *The Ethical Impact of Darwinism.* 3d ed. New York: Charles Scribner's Sons, 1903. (First edition, 1887.)

Seanor, Douglas, and N. Fotion, eds. *Hare and Critics: Essays on "Moral Thinking."* Oxford: Oxford University Press, 1988.

Searle, G. R. *Eugenics and Politics in Britain, 1900–1914.* Leiden: Nordhoff, 1976.

Searle, John. *Minds, Brains and Science.* Cambridge: Harvard University Press, 1984.

Segerstrale, Ullica. "Colleagues in Conflict: An 'In Vivo' Analysis of the Sociobiology Controversy." *Biology and Philosophy* 1, no. 1 (1986): 53–87.

Sellars, Wilfred, and John Hospers, eds. *Readings in Ethical Theory.* New York: Appleton-Century-Crofts, 1952.

Sidgwick, Arthur S., and Eleanor Mildred S. Sidgwick, eds. *Henry Sidgwick: A Memoir.* London: Macmillan, 1906.

Sidgwick, Henry. *Lectures on the Ethics of T. H. Green, Mr. Herbert Spencer, and J. Martineau.* London: Macmillan, 1902.

———. *The Methods of Ethics.* 7th ed. London: Macmillan, 1907. (First edition, 1874.)

———. "The Theory of Evolution in Its Application to Practice." *Mind* 1, no. 1 (1876): 52–67.

Simon, W. M. *European Positivism in the Nineteenth Century.* Ithaca: Cornell University Press, 1963.

Simpson, George Gaylord. *Biology and Man.* New York: Harcourt, Brace and World, 1964.

———. *The Meaning of Evolution: A Study of the History of Life and of Its Significance for Man.* New Haven: Yale University Press, 1949.

Simpson, James Y. *The Spiritual Interpretation of Nature.* London: Hodder and Stoughton, 1912.

Smith, Adam. *The Theory of Moral Sentiments.* In *The Works of Adam Smith.* Vol. 1. Aalen: Otto Zeller, 1963. Reprint of London: T. Cadell, 1811–1812.

Smith, George Allen. *The Life of Henry Drummond.* London: Hodder and Stoughton, 1899.

Smith, Goldwin. "Evolutionary Ethics and Christianity." *Contemporary Review* 44 (1883): 789–811.

———. "Has Science Yet Found a New Basis for Morality?" *Contemporary Review* 41 (1882): 335–358.

Sorell, Tom. *Recent Tendencies in Ethics: Three Lectures to Clergy Given at Cambridge.* Edinburgh: William Blackwood and Sons, 1904.

———. *Scientism: Philosophy and the Infatuation with Science.* London: Routledge, 1991.

Sorley, William Ritchie. *The Ethics of Naturalism: A Criticism.* Edinburgh: William Blackwood and Sons, 1904. (First edition, 1885.)

————. *Moral Values and the Idea of God.* Cambridge: Cambridge University Press, 1918.

Spencer, Herbert. *An Autobiography.* 2 vols. New York: D. Appleton, 1904.

————. *The Data of Ethics.* New York: Hurst and Co., n.d. (First edition, 1879.)

————. *Education: Intellectual, Moral, and Physical.* New York: D. Appleton, 1886. (First edition, 1861.)

————. *Essays: Scientific, Political, and Speculative.* 3 vols. New York: D. Appleton, 1899. (First edition, 1878.)

————. *First Principles.* 4th ed. New York: D. Appleton, 1898. (First edition, 1860–1862.)

————. *Principles of Ethics.* 2 vols. New York: D. Appleton, 1896. (First edition, 1879–1893.)

————. *Principles of Psychology.* 2 vols. New York: D. Appleton, 1881. (First edition, 1855.)

————. *Principles of Sociology.* 3 vols. New York: D. Appleton, 1897–1898. (First edition, 1876–1897.)

————. *Social Statics; or The Conditions Essential to Human Happiness Specified, and the First of Them Developed.* New York: D. Appleton, 1888. (First edition, 1851.)

————. *Various Fragments.* New York: D. Appleton, 1898. (First edition, 1897.)

Stanley, Oma. "T. H. Huxley's Treatment of 'Nature.'" *Journal of the History of Ideas* 18 (1957): 120–127.

Stent, Gunther S., ed. *Morality as a Biological Phenomenon: The Presuppositions of Sociobiological Research.* Berkeley, Los Angeles, and London: University of California Press, 1978.

Stephen, Leslie. "Darwinism and Divinity." *Fraser's Magazine* 85 (1872): 545–561.

————. "Ethics and the Struggle for Existence." *Contemporary Review* 64 (1893): 157–170.

————. *The Science of Ethics.* 2d ed. London: Smith, Elder, 1907. (First edition, 1882.)

Stephens, Lester. *Joseph Le Conte: Gentle Prophet of Evolution.* Baton Rouge: Louisiana State University Press, 1982.

Stern, Bernhard J. *Lewis Henry Morgan: Social Evolutionist.* Chicago: University of Chicago Press, 1931.

Stevenson, Charles L. *Ethics and Language.* New Haven: Yale University Press, 1944.

Stocking, George W., Jr. *Race, Culture, and Evolution: Essays in the History of Anthropology.* New York: Free Press, 1968.

————. *Victorian Anthropology.* New York: Free Press, 1987.

Stocking, George W., Jr., ed. *The Shaping of American Anthropology 1883–1911: A Franz Boas Reader*. New York: Basic Books, 1974.

Stone, Lawrence, ed. *The University and Society*. Princeton: Princeton University Press, 1974.

Storer, Morris, ed. *Humanist Ethics: Dialogue on Basics*. Buffalo: Prometheus Books, 1980.

Stout, Jeffrey. *The Flight from Authority: Religion, Morality, and the Quest for Autonomy*. Notre Dame: University of Notre Dame Press, 1981.

Stroh, Guy W. *American Ethical Thought*. Chicago: Nelson-Hall, 1979.

Suckiel, Ellen Kappy. *The Pragmatic Philosophy of William James*. Notre Dame: University of Notre Dame Press, 1982.

Sutherland, Alexander. *The Origin and Growth of the Moral Instinct*. 2 vols. London: Longmans, Green, 1898.

Symondson, Anthony, ed. *The Victorian Crisis of Faith*. London: Society for the Promotion of Christian Knowledge, 1970.

Tännsjö, Torbjörn. *Moral Realism*. Savage, Md.: Rowman and Littlefield, 1990.

Tansley, A. G. *The New Psychology and Its Relation to Life*. London: George Allen and Unwin, 1920.

Taylor, H. Terry. "William Ernest Castle, American Geneticist: A Case-Study in the Impact of the Mendelian Research Program." M.A. thesis, Oregon State University, 1973.

Teich, Mikuláš. "The Unmastered Past of Human Genetics." In Mikuláš Teich and Roy Porter, eds., *Fin de Siècle and Its Legacy*. Cambridge: Cambridge University Press, 1990. Pp. 296–324.

Teich, Mikuláš, and Robert Young, eds. *Changing Perspectives in the History of Science: Essays in Honour of Joseph Needham*. London: Heinemann, 1973.

Tennant, Neil. "Evolutionary versus Evolved Ethics." *Philosophy* 58, no. 225 (1983): 289–302.

Thayer, H. S. *Meaning and Action: A Critical History of Pragmatism*. Indianapolis: Bobbs-Merrill, 1968.

Thomas, Geoffrey. *The Moral Philosophy of T. H. Green*. Oxford: Oxford University Press, 1987.

Thomas, Laurence. "Biological Moralism." *Biology and Philosophy* 1, no. 3 (1986): 316–325.

Thomas, Malcolm. *Responses to Industrialism: The British Experience 1780–1850*. Newton Abbot: David and Charles, 1976.

Thorpe, W. H. *Science, Man and Morals*. London: Methuen, 1965.

Toulmin, Stephen. "Contemporary Scientific Mythology." In Stephen Toulmin, Ronald Hepburn, and Alasdair MacIntyre, eds., *Metaphysical Beliefs*. New York: Schocken Books, 1970. Pp. 3–71. (First edition, 1957.)

————. *An Examination of the Place of Reason in Ethics.* Cambridge: Cambridge University Press, 1968.

Trigg, Roger. "Evolutionary Ethics." *Biology and Philosophy* 1, no. 3 (1986): 325–335.

Trivers, Robert. "The Evolution of Reciprocal Altruism." *Quarterly Review of Biology* 46, no. 4 (1971): 35–75.

Trotter, Wilfred. *Instincts of the Herd in Peace and War, 1916–1919.* London: T. Fisher Unwin, 1923. (First edition, 1916.)

Tufts, James H. "Darwin and Evolutionary Ethics." *Psychology Review,* n.s., 16, no. 3 (1909): 195–206.

Turner, Frank Miller. *Between Science and Religion: The Reaction to Scientific Naturalism in Late Victorian England.* New Haven: Yale University Press, 1974.

Turner, James. *Without God, Without Creed: The Origins of Unbelief in America.* Baltimore: Johns Hopkins University Press, 1985.

Veysey, Laurence R. *The Emergence of the American University.* Chicago: Chicago University Press, 1965.

Vogeler, Martha S. *Frederic Harrison: The Vocations of a Positivist.* Oxford: Oxford University Press, 1984.

Von Arx, Jeffrey Paul. *Progress and Pessimism: Religion, Politics, and History in Late Nineteenth Century Britain.* Cambridge: Harvard University Press, 1985.

Voorzanger, Bart. "No Norms and No Nature—The Moral Relevance of Evolutionary Biology." *Biology and Philosophy* 3 (1987): 253–270.

Waddington, C. H. *The Ethical Animal.* London: George Allen and Unwin, 1960.

Waddington, C. H., ed. *Science and Ethics.* London: George Allen and Unwin, 1942.

Wallace, Alfred Russel. *Contributions to the Theory of Natural Selection.* 2d ed. London: Macmillan, 1891. (First edition, 1870.)

————. *Darwinism: An Exposition of the Theory of Natural Selection with Some of Its Applications.* London: Macmillan, 1889.

————. *Miracles and Modern Spiritualism.* Rev. ed. London: George Redway, 1896. (First edition, 1874.)

————. *My Life: A Record of Events and Opinions.* 2 vols. London: Chapman and Hall, 1905.

————. *Natural Selection and Tropical Nature: Essays on Descriptive and Theoretical Biology.* London: Macmillan, 1891.

————. "The Origin of Human Races and the Antiquity of Man Deduced from the Theory of 'Natural Selection.'" *Journal of the Anthropological Society of London* 2 (1864): clviii–clxx.

————. *Social Environment and Moral Progress.* New York: Cassell, 1913.

————. *The World of Life: A Manifestation of Creative Power, Directive Mind and Ultimate Purpose.* New York: Moffat, Yard, 1911.

Ward, Mrs. Humphry [Mary Augusta]. *Robert Elsmere.* London: Smith and Elder, 1888.

Warnock, G. J. *English Philosophy Since 1900.* 3d ed. Oxford: Oxford University Press, 1978.

————. *Morality and Language.* Oxford: Basil Blackwell, 1983.

————. *The Object of Morality.* London: Methuen, 1971.

Warnock, Mary. *Ethics since 1900.* 3d ed. Oxford: Oxford University Press, 1978.

Weber, Gay. "Science and Society in Nineteenth-Century Anthropology. *History of Science* 11 (1974): 260–283.

Webster, Charles, ed. *Biology, Medicine and Society 1840–1940.* Cambridge: Cambridge University Press, 1981.

Wedgwood, Julia. "Ethics and Science." *Contemporary Review* 72 (1897): 219–233.

Weiner, Philip. *Evolution and the Founders of Pragmatism.* Cambridge: Harvard University Press, 1949.

Weinstein, Michael A. *Unity and Variety in the Philosophy of Samuel Alexander.* West Lafayette: Purdue University Press, 1984.

Weir, James, Jr. "The Effect of Female Suffrage on Posterity." *American Naturalist* 29, no. 345 (1895): 815–825.

Westbrook, Robert B. *John Dewey and American Democracy.* Ithaca: Cornell University Press, 1991.

Westermarck, Edward. *Ethical Relativity.* London: Kegan Paul, Trench, Trubner, 1932.

————. *The Origin and Development of Moral Ideas.* 2 vols. London: Macmillan, 1906–1908.

Wheeler, William Morton. *Emergent Evolution and the Development of Societies.* New York: W. W. Norton, 1928.

White, Morton. *Science and Sentiment in America: Philosophical Thought from Jonathan Edwards to John Dewey.* Oxford: Oxford University Press, 1972.

Whitman, Charles Otis. *Posthumous Works.* 3 vols. Edited by Oscar Riddle. Washington, D.C.: Carnegie Institution, 1919.

Whorton, James C. *Crusaders for Fitness: The History of American Health Reformers.* Princeton: Princeton University Press, 1982.

Wilde, Norman. "The Meaning of Evolution in Ethics." *International Journal of Ethics* 19, no. 3 (1909): 265–283.

Wildt, Sister Carol Marie. "Julian Huxley's Conception of Evolutionary Progress." Ph.D. dissertation, Saint Louis University, 1973.

Williams, Bernard. *Ethics and the Limits of Philosophy.* Cambridge: Harvard University Press, 1985.

————. "Evolution, Ethics, and the Representation Problem." In D. S. Bendall, ed., *Evolution from Molecules to Man*. Cambridge: Cambridge University Press, 1983. Pp. 555–566.

————. "Evolutionary Theory: Epistemology and Ethics." In Alan Grafen, ed., *Evolution and Its Influence*. Oxford: Oxford University Press, 1989. Pp. 93–106.

————. *Moral Luck: Philosophical Papers*. Cambridge: Cambridge University Press, 1981.

————. *Morality: An Introduction to Ethics*. New York: Harper and Row, 1972.

Williams, Cora M. *A Review of the Systems of Ethics Founded on the Theory of Evolution*. London: Macmillan, 1893.

Williams, George C. *Adaptation and Natural Selection: A Critique of Some Current Evolutionary Thought*. Princeton: Princeton University Press, 1966.

Williams, Patricia. "Evolved Ethics Re-Examined: The Theory of Robert J. Richards." *Biology and Philosophy* 5 (1990): 451–457.

Williams, Raymond. *Culture and Society 1780–1950*. New York: Columbia University Press, 1960.

Wilson, Daniel J. *Science, Community, and the Transformation of American Philosophy, 1860–1930*. Chicago: University of Chicago Press, 1990.

Wilson, David Sloan. "On the Relationship Between Evolutionary and Psychological Definitions of Altruism and Selfishness." *Biology and Philosophy* 7 (1992): 61–68.

Wilson, Edward O. "Introduction: What Is Sociobiology?" In Michael Gregory, Anita Silvers, and Diane Sutch, eds., *Sociobiology and Human Nature*. San Francisco: Jossey-Bass, 1978.

————. *On Human Nature*. Cambridge: Harvard University Press, 1978.

————. *Sociobiology: The New Synthesis*. Cambridge: Harvard University Press, 1975.

Wilson, James Q. *The Moral Sense*. New York: Free Press, 1993.

Wiltshire, David. *The Social and Political Thought of Herbert Spencer*. Oxford: Oxford University Press, 1978.

Winston, George R. *John Fiske*. New York: Twayne, 1972.

Wollheim, Richard. *F. H. Bradley*. London: Penguin, 1969.

Wolstenholme, Gordon, ed. *Man and His Future: A Ciba Foundation Volume*. London: J. and A. Churchill, 1963.

Wright, Chauncey. "*Contributions to the Theory of Natural Selection: A Series of Essays*. By Alfred Russel Wallace, Author of 'The Malay Archipelago,' etc., etc." *North American Review* 111 (1870): 282–311.

————. "Evolution of Self-Consciousness." *North American Review* 116, no. 239 (1873): 245–310.

————. "German Darwinism." *Nation* 21, no. 532 (1875): 168–170.

Wright, H. W. "Evolution and Ethical Method." *International Journal of Ethics* 16, no. 1 (1905): 59–68.

Wynne-Edwards, V. C. *Animal Dispersion in Relation to Social Behavior.* Edinburgh: Oliver and Boyd, 1962.

Yolton, John W. *Thinking Matter: Materialism in Eighteenth-Century Britain.* Minneapolis: University of Minnesota Press, 1984.

Youmans, Edward Livingston, ed. *Herbert Spencer on the Americans and the Americans on Herbert Spencer: Being a Full Report of His Interview, and of the Proceedings of the Farewell Banquet of Nov. 11, 1882.* New York: D. Appleton, 1883.

Zernel, John. "John Wesley Powell: Science and Reform in a Positive Context." Ph.D. dissertation, Oregon State University, 1983.

Zink, David. *Leslie Stephen.* New York: Twayne, 1972.

Zumback, Clark. *The Transcendental Science: Kant's Conception of Biological Method.* The Hague: Nijhoff, 1984.

Index

Designer:	U.C. Press Staff
Compositor:	Prestige Typography
Text:	10/12 Sabon
Display:	Sabon
Printer:	Maple-Vail Book Manufacturing Group
Binder:	Maple-Vail Book Manufacturing Group